Conference Proceedings of the Society for Experimental Mechanics Series

Series Editor

Kristin B. Zimmerman, Ph.D.
Society for Experimental Mechanics, Inc.,
Bethel, CT, USA

More information about this series at http://www.springer.com/series/8922

Antonio Baldi • Simon Quinn • Xavier Balandraud
Janice M. Dulieu-Barton • Sven Bossuyt
Editors

Residual Stress, Thermomechanics & Infrared Imaging, Hybrid Techniques and Inverse Problems, Volume 7

Proceedings of the 2018 Annual Conference on Experimental and Applied Mechanics

Springer

Editors
Antonio Baldi
Mechanical Engg, Chem & Materials
Faculty of Engg & Architecture
Cagliari, Italy

Xavier Balandraud
Université Clermont Auvergne
CNRS, SIGMA Clermont
Institut Pascal
Clermont-Ferrand, France

Sven Bossuyt
Aalto University
Aalto, Finland

Simon Quinn
Envir Enterprise Zone, Building 15
EDMC Stores, Rifi, Eng &
Southampton, UK

Janice M. Dulieu-Barton
Engineering Materials Research Group
Faculty of Engineering and the Environment
University of Southampton
Southhampton, UK

ISSN 2191-5644 ISSN 2191-5652 (electronic)
Conference Proceedings of the Society for Experimental Mechanics Series
ISBN 978-3-030-06970-4 ISBN 978-3-319-95074-7 (eBook)
https://doi.org/10.1007/978-3-319-95074-7

This Springer imprint is published by the registered company Springer Nature Switzerland AG
The registered company address is: Gewerbestrasse 11, 6330 Cham, Switzerland

Preface

Residual Stress, Thermomechanics & Infrared Imaging, Hybrid Techniques and Inverse Problems represents one of eight volumes of technical papers presented at the 2018 SEM Annual Conference and Exposition on Experimental and Applied Mechanics organized by the Society for Experimental Mechanics and held in Greenville, SC, June 4–7, 2018. The complete Proceedings also includes volumes on: *Dynamic Behavior of Materials*; *Challenges in Mechanics of Time-Dependent Materials*; *Advancement of Optical Methods & Digital Image Correlation in Experimental Mechanics*; *Mechanics of Biological Systems & Micro- and Nanomechanics*; *Mechanics of Composite, Hybrid & Multifunctional Materials*; *Fracture, Fatigue, Failure and Damage Evolution*; and *Mechanics of Additive and Advanced Manufacturing.*

Each collection presents early findings from experimental and computational investigations on an important area within experimental mechanics; residual stress, thermomechanics and infrared imaging, hybrid techniques, and inverse problems being a few of these areas.

Residual stresses are self-balanced stress fields induced during most material processing procedures, for example, welding/joining, casting, thermal conditioning, and forming. Their hidden character often causes them to be underrated or overlooked. However, they profoundly influence structural design and substantially affect strength, fatigue life, and dimensional stability. Thus, they must be taken seriously and included in practical applications.

In recent years, the applications of infrared imaging techniques to the mechanics of materials and structures have grown considerably. The expansion is marked by the increased spatial and temporal resolution of the infrared detectors, faster processing times, much greater temperature resolution, and specific image processing. The improved sensitivity and more reliable temperature calibrations of the devices have meant that more accurate data can be obtained than were previously available.

Advances in inverse identification have been coupled with optical methods that provide surface deformation measurements and volumetric measurements of materials. In particular, inverse methodology was developed to more fully use the dense spatial data provided by optical methods to identify mechanical constitutive parameters of materials. Since its beginnings during the 1980s, creativity in inverse methods has led to applications in a wide range of materials, with many different constitutive relationships, across material heterogeneous interfaces. Complex test fixtures have been implemented to produce the necessary strain fields for identification. Force reconstruction has been developed for high strain rate testing. As developments in optical methods improve for both very large and very small length scales, applications of inverse identification have expanded to include geological and atomistic events. Researchers have used in situ 3D imaging to examine microscale expansion and contraction and used inverse methodologies to quantify constitutive property changes in biological materials.

Southampton, UK	Janice M. Dulieu-Barton
Cagliari, Italy	Antonio Baldi
Aalto, Finland	Sven Bossuyt
Southampton, UK	Simon Quinn
Clermont-Ferrand, France	Xavier Balandraud

Contents

Chapter 1
Comparison Between 0D and 1D Heat Source Reconstruction for Fatigue Characterization

Pawarut Jongchansitto, Corentin Douellou, Itthichai Preechawuttipong, and Xavier Balandraud

Abstract Material fatigue damage is associated with heat production leading to material self-heating. In this context, measuring temperature fields on a specimen's surface by infrared thermography is useful to analyze the fatigue response of the tested material. Calorific information can be also obtained by reconstructing the heat power density (heat sources) at the origin of the temperature changes. In particular, mechanical dissipation due to fatigue damage can be determined from the thermal response using specific temperature acquisition conditions. The processing is based the heat diffusion equation, whose different formulations have been proposed in the literature to perform heat source reconstruction. The present study compares two approaches for homogeneous fatigue tests, namely the zero-dimensional (0D) and one-dimensional (1D) approaches. The error generated by the 0D approach (compared to the 1D approach) was first determined from a model. For comparison purposes, experimental tests were performed on a copper specimen. Consequences of using the 0D processing for fatigue analysis are discussed.

Keywords Infrared thermography · Fatigue · Heat source · Mechanical dissipation · 0D approach

1.1 Introduction

Material fatigue is generally associated with a production of *mechanical dissipation* or *intrinsic dissipation*, leading to a "self-heating". The heat sources due to fatigue can be calculated from the temperatures measured on the surface of a specimen using infrared (IR) thermography. By "heat sources", we mean the heat power density (in $W \cdot m^{-3}$) which is produced or absorbed by the material due to a change in the material thermomechanical state. When heat sources are spatially homogeneous in the tested specimen, processing can be performed using a macroscopic approach, also named zero-dimensional (0D) approach. The latter consists in using the temperature changes averaged over the whole specimen's measurement zone to calculate the heat sources [1–6]. This approach is based on the fact that in the case of spatially homogeneous heat sources, the heat exchanged by conduction with the jaws of the testing machine is nearly proportional to the mean temperature change [7]. The 0D processing considers a "global" heat exchange with the outside of the tested specimen (jaws and ambient air), which is proportional to the mean temperature change. The present study analyzes the error generated by the 0D approach in the assessment of mechanical dissipation during a fatigue test. This error can be defined with respect to a one-dimensional (1D) approach, which is applicable to longitudinal specimens subjected to uniaxial loading [8 13]. Heat exchanges by conduction in the specimen and by convection with the ambient air are separately taken into account in the 1D approach.

The study was performed in two steps. First, a model was proposed to estimate the error generated by the 0D approach in the assessment of mechanical dissipation due to fatigue damage. Second, tests were performed on a specimen made of pure copper.

P. Jongchansitto · I. Preechawuttipong
Chiang Mai University, Faculty of Engineering, Department of Mechanical Engineering, Muang District, Chiang Mai, Thailand

C. Douellou · X. Balandraud (✉)
Université Clermont Auvergne, CNRS, SIGMA Clermont, Institut Pascal, Clermont-Ferrand, France
e-mail: xavier.balandraud@sigma-clermont.fr

A. Baldi et al. (eds.), *Residual Stress, Thermomechanics & Infrared Imaging, Hybrid Techniques and Inverse Problems, Volume 7*,
Conference Proceedings of the Society for Experimental Mechanics Series, https://doi.org/10.1007/978-3-319-95074-7_1

1.2 Preliminary Analysis from a Model

Let us consider a longitudinal specimen with constant cross-section and placed in the jaws of a uniaxial testing machine: see Fig. 1.1. The temperature change $\theta(z, t)$ due to mechanical loading is given by the following 1D expression of the heat diffusion equation:

$$\rho C \left(\frac{\partial \theta}{\partial t} + \frac{\theta}{\tau} \right) - \lambda \frac{\partial^2 \theta}{\partial z^2} = s \tag{1.1}$$

where s is the variation over time t in the heat source. The density ρ, specific heat C and thermal conductivity coefficient λ are assumed to be homogeneous in the specimen. τ is a time constant characterizing the heat exchanges by convection with ambient air. The specimen is clamped in the jaws of the machine. Considering that specimen and jaws are both metallic, the following boundary conditions are assumed:

$$\theta \left(\pm \frac{L}{2}, t \right) = 0 \tag{1.2}$$

where L is the clear distance between the two jaws and $z = 0$ the mid-point of the specimen.

The heat source $s(t)$ corresponds to the heat power density produced or absorbed by the material itself. Heat sources due to thermoelastic coupling are negative when the stress increases, and positive when the stress decreases. Heat sources due to irreversible phenomena are always positive. They are named *mechanical dissipation* or *intrinsic dissipation*. Let us consider a fatigue test at constant frequency, maximum force and force ratio. The heat due to thermoelastic coupling is null over a mechanical cycle. As a consequence, it is possible to write an averaged version of Eq. (1.1) in steady-state regime as follows [14]:

$$\rho C \frac{\theta}{\tau} - \lambda \frac{\partial^2 \theta}{\partial z^2} = d_1 \tag{1.3}$$

where d_1 is the mean mechanical dissipation per cycle, assumed to be constant along the test. The solution to Eqs. (1.2) and (1.3) writes as follows [15]:

$$\theta\,(z, \text{steady}) = \frac{d_1}{\lambda\, \beta^2} \left[1 - \frac{\cosh\,(\beta\, z)}{\cosh\left(\beta\, \frac{L}{2}\right)} \right] \tag{1.4}$$

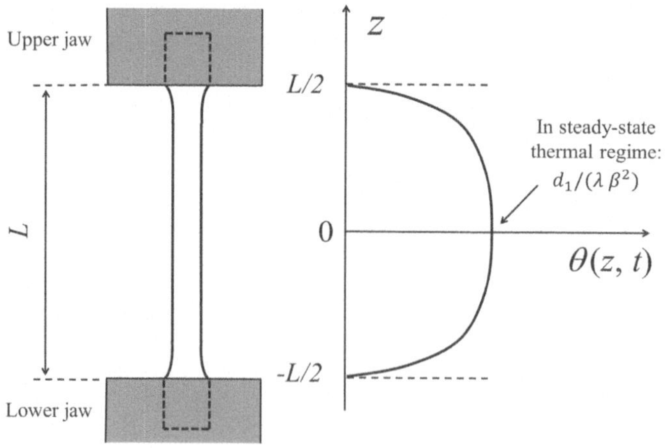

Fig. 1.1 Schematic view of a metallic specimen subjected to uniaxial fatigue loading

where

$$\beta = \sqrt{\frac{\rho\, C}{\lambda\, \tau}} \tag{1.5}$$

The temperature change to be considered for the 0D approach is defined by:

$$\Theta_{0D}\,(\text{steady}) = \frac{1}{L}\int_{-L/2}^{L/2} \theta\,(z,\,\text{steady})\;dz \tag{1.6}$$

Then it gives from Eqs. (1.4) and (1.6):

$$\Theta_{0D}\,(\text{steady}) = \frac{d_1}{\lambda\,\beta^2}\left[1 - \frac{2}{\beta\, L}\,\tanh\left(\beta\,\frac{L}{2}\right)\right] \tag{1.7}$$

A numerical model was developed to find an empirical expression for the global time constant τ_{0D} characterizing the global heat exchanges with the outside of the specimen (by contact with the jaws and by convection with ambient air). An implicit Euler scheme of Eq. (1.1) was implemented to obtain synthetic values of $\theta(z,t)$ during a natural return to ambient temperature ($s = 0$) after homogeneous heating. Numerical results showed that the value of τ_{0D} can be written as follows for pure copper [14]:

$$\tau_{0D} \approx 61.7\; L^{1.65}\; \tau^{0.373} \tag{1.8}$$

with τ_{0D} and τ are expressed in [s], and L in [m]. The mechanical dissipation can be then estimated by using the 0D version of the heat equation in the steady-state regime:

$$(d_1)_{0D} = \rho C \frac{\Theta_{0D}\,(\text{steady})}{\tau_{0D}} \tag{1.9}$$

It gives from Eqs. (1.7), (1.8) and (1.9):

$$(d_1)_{0D} = \frac{1 - \frac{2}{\beta\, L}\,\tanh\left(\beta\,\frac{L}{2}\right)}{61.7\;\, L^{1.65}\;\tau^{-0.627}} \times d_1 \tag{1.10}$$

The relative error in per-cent generated by the 0D approach in the evaluation of mechanical dissipation can be then expressed as follows:

$$E\% = \left[\frac{1 - \frac{2}{\beta\, L}\,\tanh\left(\beta\,\frac{L}{2}\right)}{61.7\;\, L^{1.65}\;\tau^{-0.627}} - 1\right] \times 100 \tag{1.11}$$

1.3 Experimental Analysis

Fatigue test was performed on a specimen made of pure copper. Figure 1.2a, b present the specimen geometry. Two reference samples were also employed to track variations in the specimen's environment [14]. The references were clamped in the jaws of the machine. They were made of the same material as the tested specimen. The test was performed using a hydraulic ±15 kN MTS testing machine, see Fig. 1.1. To maximize the thermal emissivity of the specimen and the two references, surfaces were painted in a matte black color. Black curtains were also placed around the immediate environment to reduce reflections.

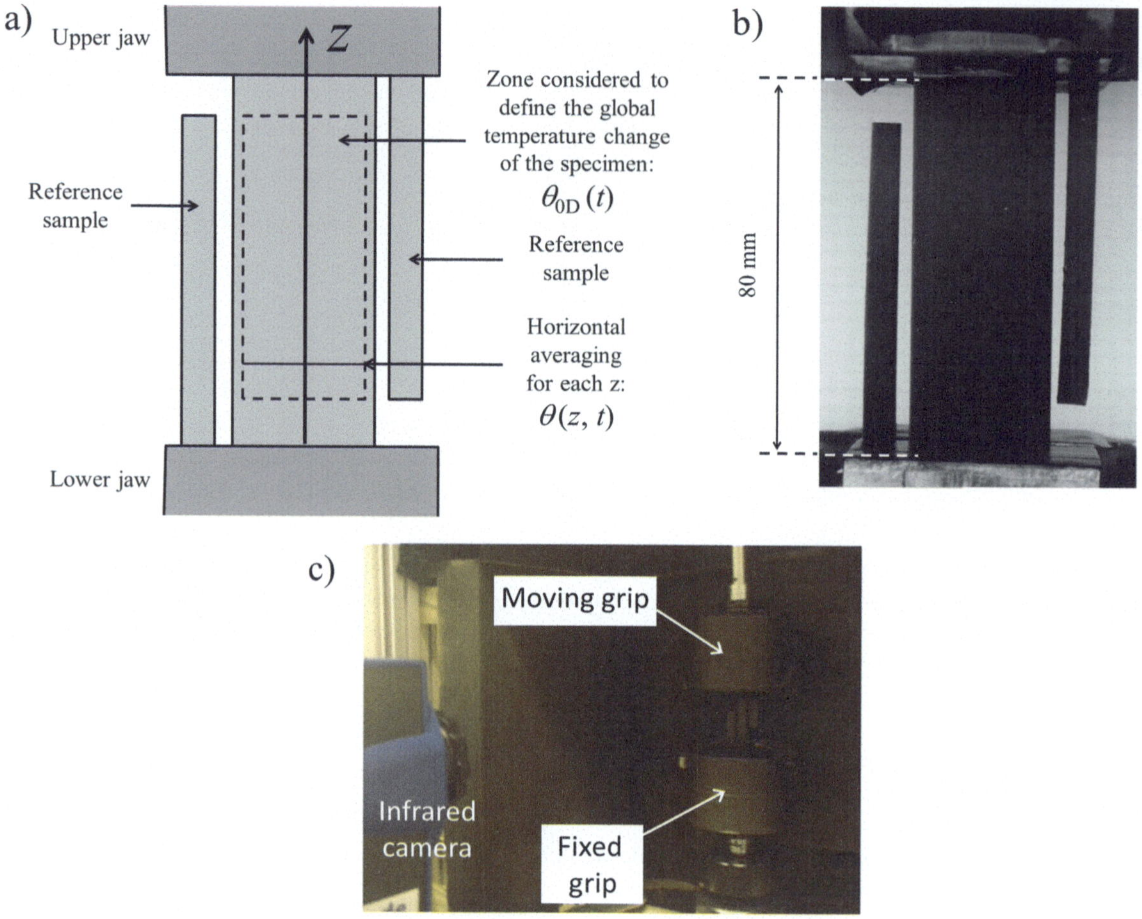

Fig. 1.2 Experimental set-up: (**a**) schematic view of the specimen with two reference samples, (**b**) specimen and reference samples painted in black to maximize the thermal emissivity, (**c**) photo of the experimental set-up

A Jade III-MWIR camera was employed to capture the temperature fields during the test. A force-controlled sinusoidal loading was applied to the specimen for 150 s. The temperature acquisition conditions were such that each recording image corresponded to an average over several mechanical cycles. As the heat due to thermoelastic coupling is null over a cycle, the heat sources that are calculated from Eq. (1.1) thus correspond to the mean mechanical dissipation. Prior to fatigue testing, the time constant τ characterizing the heat exchanges with ambient air was identified from a natural return to ambient temperature while the specimen is not clamped in the jaws of the machine (it was merely suspended in air).

The 0D version of the heat equation writes as follows

$$\rho C\left(\frac{\partial \Theta_{0D}}{\partial t}+\frac{\Theta_{0D}}{\tau_{0D}}\right)=s \tag{1.12}$$

The calculation of $(d_1)_{0D}$ thus requires the knowledge of the time constant τ_{0D} characterizing the global heat exchanges with the outside of the specimen (ambient air and jaws of the machine). In practice, τ_{0D} was estimated from a natural return to ambient temperature while the specimen is clamped in the jaws.

Figure 1.3 presents a temperature change profile in the thermal steady-state regime of a cyclic loading. The mean temperature change was obtained by spatial averaging over the specimen surface. This quantity enabled us to calculate the mechanical dissipation $(d_1)_{0D}$ from Eq. (1.9) with the experimental data. Equation (1.3) was used to determine the mechanical dissipation d_1 from the 1D approach. In practice, calculation was done at the mid-point of the specimen, i.e. at $z = 0$: see red curve in Fig. 1.3, employed to estimate the two terms in the left-hand side of Eq. (1.3). Numerical application led to a relative error $E\%$ of -8% in the estimation of the mechanical dissipation by the 0D approach compared to the 1D approach. The theoretical value from Eq. (1.11) is of -15%. This quantity actually strongly changes as a function of the

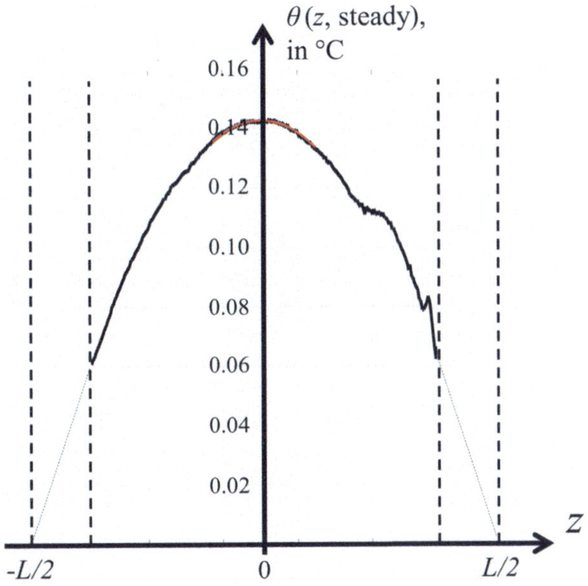

Fig. 1.3 Fatigue test of copper: temperature change profile in steady-state regime

specimen's length L and of the time constant τ characterizing the heat exchanges by convection with air [15]. Considering all the modeling hypotheses, it can be said that the experimental and simulated results are here in fair agreement. It must be noted that, although the relative error $E\%$ may seem small, it is important for quantitative fatigue analysis.

1.4 Conclusion

The present study aimed at analyzing the error generated by the 0D approach in the evaluation of mechanical dissipation due to fatigue damage. A copper specimen was subjected to cyclic loading while temperatures were measured by IR thermography. A model was also developed to provide a theoretical expression of the error generated by the 0D processing. The results showed that this error may not be negligible. It can be concluded that the 0D approach is a powerful and quite simple tool for thermomechanical analysis of materials, at least when heat sources are spatially homogeneous, but attention must be paid on its application for quantitative fatigue analysis.

Acknowledgements The authors gratefully acknowledge the French Embassy in Thailand and Campus France for their support during this research (PHC SIAM 2018, Project 40710SE). The authors also gratefully acknowledge the Ministere de l'Europe et des Affaires Etrangeres (MEAE) and the Ministere de l'Enseignement superieur, de la Recherche et de l'Innovation (MESRI) in France, as well as the office of the Higher Education Commission (OHEC) of the Ministry of Education in Thailand.

References

1. Boulanger, T., Chrysochoos, A., Mabru, C., Galtier, A.: Calorimetric analysis of dissipative and thermoelastic effects associated with the fatigue behavior of steels. Int. J. Fatigue. **26**, 221–229 (2004)
2. Giancane, S., Chrysochoos, A., Dattoma, V., Wattrisse, B.: Deformation and dissipated energies for high cycle fatigue of 2024-T3 aluminium alloy. Theor. Appl. Fract. Mec. **52**, 117–121 (2009)
3. Samaca Martinez, J.R., Le Cam, J.B., Balandraud, X., Toussaint, E., Caillard, J.: Filler effects on the thermomechanical response of stretched rubbers. Polym. Test. **32**, 835–841 (2013)
4. Benaarbia, A., Chrysochoos, A., Robert, G.: Kinetics of stored and dissipated energies associated with cyclic loadings of dry polyamide 6.6 specimens. Polym. Test. **34**, 155–167 (2014)
5. Samaca Martinez, J.R., Le Cam, J.B., Balandraud, X., Toussaint, E., Caillard, J.: New elements concerning the Mullins effect: a thermomechanical analysis. Eur. Polym. J. **55**, 98–107 (2014)
6. Wang, X.G., Crupi, V., Jiang, C., Feng, E.S., Guglielmino, E., Wang, C.S.: Energy-based approach for fatigue life prediction of pure copper. Int. J. Fatigue. **104**, 243–250 (2017)
7. Chrysochoos, A., Louche, H.: An infrared image processing to analyse the calorific effects accompanying strain localization. Int. J. Eng. Sci. **38**, 1759–1788 (2000)

8. Balandraud, X., Chrysochoos, A., Leclercq, S., Peyroux, R.: Influence of the thermomechanical coupling on the propagation of a phase change front. Comptes-Rendus de l'Académie des Sciences Série II Fascicule B Mécanique. **329**, 621–626 (2001)
9. Wattrisse, B., Muracciole, J.M., Chrysochoos, A.: Thermomechanical effects accompanying the localized necking of semi-crystalline polymers. Int. J. Therm. Sci. **41**, 422–427 (2002)
10. Connesson, N., Maquin, F., Pierron, F.: Experimental energy balance during the first cycles of cyclically loaded specimens under the conventional yield stress. Exp. Mech. **51**, 23–44 (2011)
11. Blanche, A., Chrysochoos, A., Ranc, N., Favier, V.: Dissipation assessments during dynamic very high cycle fatigue tests. Exp. Mech. **55**, 699–709 (2015)
11. Ranc, N., Blanche, A., Ryckelynck, D., Chrysochoos, A.: POD preprocessing of IR thermal data to assess heat source distribution. Exp. Mech. **55**, 725–739 (2015)
13. Delobelle, V., Favier, D., Louche, H., Connesson, N.: Determination of local thermophysical properties and heat of transition from thermal fields measurement during drop calorimetric experiment. Exp. Mech. **55**, 711–723 (2015)
14. Delpueyo, D., Balandraud, X., Grédiac, M., Stanciu, S., Cimpoesu, N.: A specific device for enhanced measurement of mechanical dissipation in specimens subjected to long-term tensile tests in fatigue. Strain. **54**, e12252 (2017)
15. Jongchansitto, P., Douellou, C., Preechawuttipong, I., Balandraud, X.: Comparison between 0D and 1D approaches for mechanical dissipation measurement during fatigue tests. Strain. (2018, submitted)

Chapter 2
Analysis of the Thermomechanical Response of Granular Materials by Infrared Thermography

Pawarut Jongchansitto, Xavier Balandraud, Itthichai Preechawuttipong, Jean-Benoît Le Cam, and Pierre Garnier

Abstract Granular materials are defined as a collection of solid particles whose macroscopic mechanical behavior is governed by the interaction forces between the particles. Full-field experimental data on these materials remain few compared to numerical results, even though a wide literature deals with optical imaging (combined with digital image correlation) and photoelasticimetry (to measure shear stresses in particles made of birefringent materials). We applied infrared thermography to analyze two-dimensional granular media composed of cylinders and subjected to confined compression. We analyzed the calorific signature of the contact forces, especially by revealing mechanical dissipation in the interparticle friction zones. Moreover, two constitutive materials featuring entropic and isentropic elasticity were employed to compare distinct types of thermoelastic couplings. Couplings and mechanical dissipation were separately identified at two observation scales. The perspective of this work is the experimental analysis of soft granular media.

Keywords Granular material · Infrared thermography · Thermoelastic coupling · Friction · Entropic coupling

2.1 Introduction

Granular materials are omnipresent in our daily life as well as in many industrial fields. They are defined as a collection of solid particles whose macroscopic mechanical behavior is governed by the contact forces between the particles. They generally consist of grains with wide distributions in terms of size, shape and base material. Their mechanical behavior strongly differs from that of continuum solids, liquids, and gases [1]. Numerical simulation has been carried out to investigate the influence of parameters such as the particle shape, density, polydispersity, particle elasticity, friction, etc. [2]. Several experimental techniques are also available. For measurements in the bulk, let us cite X-ray tomography [3, 4], flash X-ray shadowgraphy [5], magnetic resonance imaging [6], radar-based sensing [7] and positron emission particle tracking [8]. For two-dimensional (2D) measurements, particle image velocimetry and digital image correlation were widely employed [9–11]. For stress measurement in the particles, photoelasticimetry is a full-field experimental technique providing the shear stresses in particles made of birefringent materials [12–16]. In the present study, we employed infrared (IR) thermography to reveal the calorific signature of the contact forces, especially in the interparticle friction zones. Moreover, two base materials were employed to compare two types of thermoelastic coupling: *isentropic* coupling (typical of metallic materials and polymers below their glass transition temperature) and *entropic coupling* (typical of rubber-like materials).

P. Jongchansitto · I. Preechawuttipong
Chiang Mai University, Faculty of Engineering, Department of Mechanical Engineering, Muang District, Chiang Mai, Thailand

X. Balandraud (✉)
Université Clermont Auvergne, CNRS, SIGMA Clermont, Institut Pascal, Clermont-Ferrand, France
e-mail: xavier.balandraud@sigma-clermont.fr

J.-B. Le Cam
Univ Rennes, CNRS, IPR (Institute de Physique de Rennes) – UMR 6251, Rennes, France

LC-DRIME, Joint Research Laboratory, Cooper Standard – Institut de Physique UMR 6251, Rennes Cedex, France

P. Garnier
PCM Technologies S.A.S., Rue René Moineau, Champtocé-sur-Loire, France

A. Baldi et al. (eds.), *Residual Stress, Thermomechanics & Infrared Imaging, Hybrid Techniques and Inverse Problems, Volume 7*, Conference Proceedings of the Society for Experimental Mechanics Series, https://doi.org/10.1007/978-3-319-95074-7_2

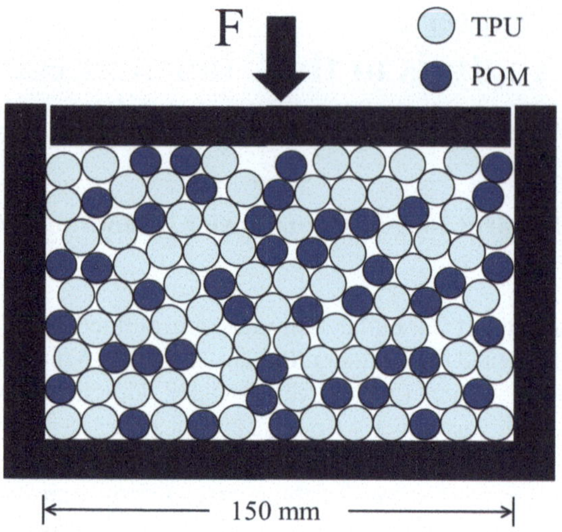

Fig. 2.1 Experimental configuration

2.2 Methodology

The experimental configuration is presented in Fig. 2.1. Cylinders made of two types of materials were employed: thermoplastic polyurethane (TPU) and polyoxymethylene (POM). These materials differ in terms of stiffness and thermoelastic coupling. TPU is much softer than POM (ratio of about 700 in tension). TPU is governed by entropic thermoelastic coupling [17, 18] while POM is governed by isentropic thermoelastic coupling [19–21] at ambient temperature and within the range of mechanical loadings applied in the study. TPU cylinders were molded by PCM Technologies S.A.S. (France). See Refs. [19–21] for the properties of POM cylinders. POM cylinders were 10.5 mm in diameter. TPU cylinders were 12.0 mm in diameter. All cylinders were 60 mm in length. The surface of each cylinder was painted in a matte black color to homogenize and maximize the thermal emissivity.

A cyclic mechanical loading was applied to the top of the granular sample using a MTS $\pm$15 kN uniaxial testing machine: force-controlled sinusoidal signal, frequency of 0.5 Hz, force ratio of 0.1, maximum compression force of -10 kN. Ambient temperature was equal to 28 °C. A Cedip Jade III-MWIR camera was employed to capture the temperature fields on the surface of the cylinders at an acquisition frequency of 150 Hz.

2.3 Global Analysis

In order to reveal irreversible phenomena occurring during the cyclic loading, we assume that the heat produced by the thermoelastic response is equal to zero over a mechanical cycle (or an integer number of cycles). The heat produced by mechanical irreversibility is always positive. The corresponding heat power density is usually named *mechanical dissipation* or *intrinsic dissipation* [22–24]. For granular materials, mechanical dissipation can be due to friction at the contacts between particles. It can be also due to the stresses inside the particles, in particular in the stress concentration zones. Figure 2.2 shows a map of temperature differences for three mechanical cycles. It can be observed that the values are close to zero except at some contacts between particles. In particular, strong thermal activity is observed along some lines inclined with respect to the loading direction. It can be checked that the higher the number of mechanical cycles considered, the higher the temperature differences due to irreversible phenomena.

In order to reveal thermoelastic couplings in the material's response during the cyclic loading, we have considered a half-cycle. The heat due to thermoelastic couplings for loading and unloading are indeed equal but with opposite signs. Figure 2.3 shows a map of temperature differences for a compression stage. It can be noted that the values are greater than those obtained in Fig. 2.2 corresponding to three entire cycles. This shows that that the amplitude of the thermoelastic effects is much greater than the thermal signature of the irreversible phenomena. It can also be observed in Fig. 2.3 that heating is not visualized only at the contacts: it is distributed inside the particles. Hotter particles are mainly made of TPU. The entropic coupling is indeed stronger than the isentropic coupling. Next section provides information at a better spatial resolution.

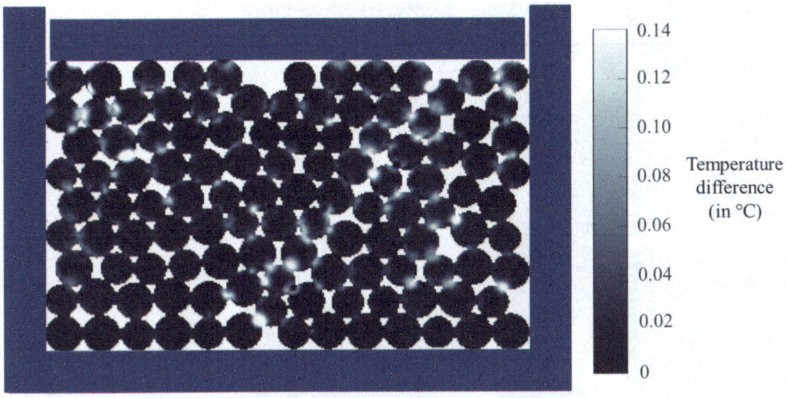

Fig. 2.2 Map of temperature differences for three mechanical cycles, revealing mechanical dissipation

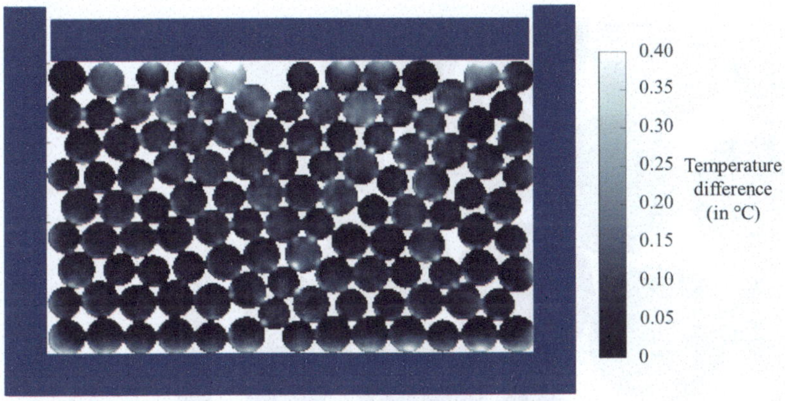

Fig. 2.3 Map of temperature differences for a half-cycle, revealing thermoelastic couplings

2.4 Local Analysis

The IR camera was then placed closer to the granular sample to obtain a better spatial resolution in a specific zone of the granular assembly. Similarly to the global analysis, the map of temperature differences for three mechanical cycles is presented in Fig. 2.4 to reveal mechanical dissipation. Hot zones are observed on both sides of the contact zones. This can be explained by the diffusion of the heat produced by the friction [25]. Heterogeneity in the intensity of the irreversible thermal response is clearly evidenced.

Similarly to the global analysis, the map of temperature differences for a half-cycle is presented in Fig. 2.5 to reveal thermoelastic coupling effects. The thermal patterns differ from those observed in Fig. 2.4 corresponding to irreversible phenomena. Indeed, the thermoelastic response depends on the type of material (POM or TPU) and on the local stress state, whereas the mechanical dissipation depends on the friction intensity at the contact. For example, it is possible to have a strong stress concentration (leading to strong thermoelastic effect) accompanied by low friction depending on the ratio between normal and transverse contact forces.

2.5 Conclusion

IR thermography is in principle applicable to any type of solid material to perform a thermomechanical analysis. However, the application to granular assemblies is difficult. The present study provides arguments for choosing constitutive materials for such analyses. It can be claimed that entropic elastic matters are excellent candidates as analogical materials for the analysis of granular media by IR thermography. Rubber-like entropic particles (such as TPU) provide strong thermoelastic responses that could be used to identify the contact force network. In addition, thermoelastic couplings and mechanical dissipation due to interparticle friction can be separately revealed. IR thermography thus provides two routes to analyze

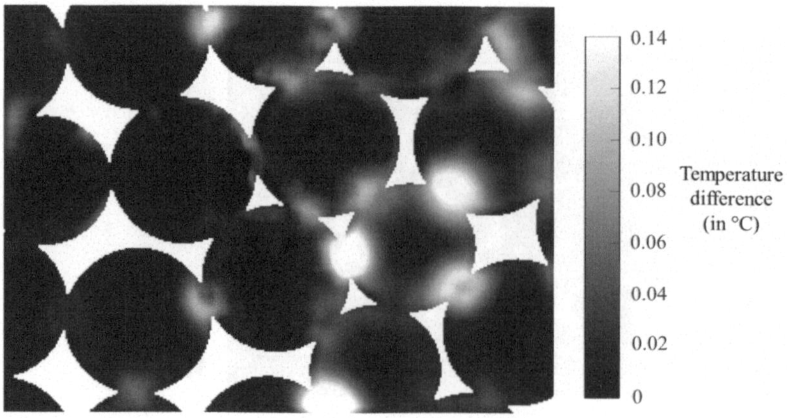

Fig. 2.4 Local analysis: map of temperature differences for three mechanical cycles, revealing mechanical dissipation

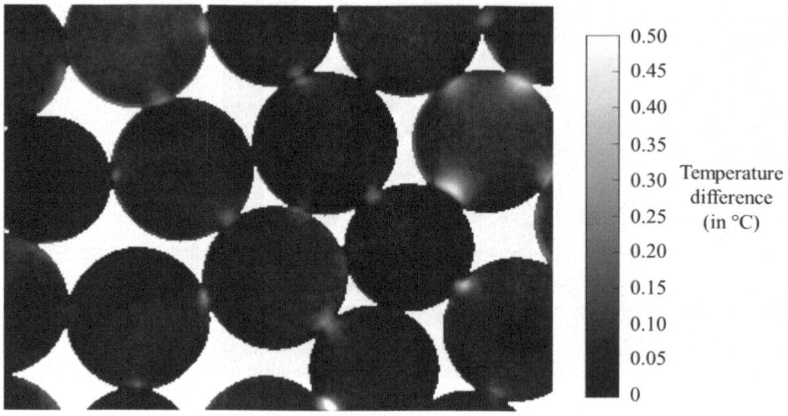

Fig. 2.5 Local analysis: map of temperature differences for a half-cycle, revealing thermoelastic couplings

granular materials. Compared to previous works performed with only isentropic materials [19–21], numerous mechanical cycles and high loading frequency are not required for the processing. Only a few mechanical cycles are necessary for the analysis: a half-cycle for revealing entropic thermoelastic coupling, and one cycle minimum for revealing irreversible phenomena. More generally, the study opens prospects for the experimental analysis of "soft" granular materials.

Acknowledgements The authors gratefully acknowledge the company PCM Technologies S.A.S., Champtocé-sur-Loire, France for their support during this research. They also gratefully acknowledge the French Embassy in Thailand and Campus France for their support during this research (PHC SIAM 2018, Project 40710SE). Finally, the authors gratefully acknowledge the Ministere de l'Europe et des Affaires Etrangeres (MEAE) and the Ministere de l'Enseignement superieur, de la Recherche et de l'Innovation (MESRI) in France, as well as the office of the Higher Education Commission (OHEC) of the Ministry of Education in Thailand.

References

1. Jaeger, H.M.: Sand, jams and jets. Phys. World. **18**, 34–39 (2005)
2. Radjai, F., Roux, J.N., Daouadji, A.: Modeling granular materials: century-long research across scales. J. Eng. Mech. **143**, 04017002 (2017)
3. Hall, S.A., Bornert, M., Desrues, J., Pannier, Y., Lenoir, N., Viggiani, G., Besuelle, P.: Discrete and continuum analysis of localised deformation in sand using X-ray μCT and volumetric digital image correlation. Geotechnique. **60**, 315–322 (2010)
4. Khalili, M.H., Brisard, S., Bornert, M., Aimedieu, P., Pereira, J.M., Roux, J.N.: Discrete digital projections correlation: a reconstruction-free method to quantify local kinematics in granular media by X-ray tomography. Exp. Mech. **57**, 819–830 (2017)
5. Park, H., Chen, W.: Two orthogonal layers of metal medium in granular materials for 3D speckle shadowgraph by flash X-ray. Exp. Mech. **52**, 1173–1177 (2012)
6. Nakagawa, M., Altobelli, S.A., Caprihan, A., Fukushima, E., Jeong, E.K.: Non-invasive measurements of granular flows by magnetic resonance imaging. Exp. Fluids. **16**, 54–60 (1993)

7. Hill, K.M., Fan, Y., Zhang, J., Van Niekerk, C., Zastrow, E., Hagness, S.C., Bernhard, J.T.: Granular segregation studies for the development of a radar-based three-dimensional sensing system. Granul. Matter. **12**, 201–207 (2010)
8. Parker, D.J., Dijkstra, A.E., Martin, T.W., Seville, J.P.K.: Positron emission particle tracking studies of spherical particle motion in rotating drums. Chem. Eng. Sci. **52**, 2011–2022 (1997)
9. Slominski, C., Niedostatkiewicz, M., Tejchman, J.: Application of particle image velocimetry (PIV) for deformation measurement during granular silo flow. Powder Technol. **173**, 1–18 (2007)
10. Hall, S.A., Wood, D.M., Ibraim, E., Viggiani, G.: Localised deformation patterning in 2D granular materials revealed by digital image correlation. Granul. Matter. **12**, 1–14 (2010)
11. Richefeu, V., Combe, G., Viggiani, G.: An experimental assessment of displacement fluctuations in a 2D granular material subjected to shear. Geotech. Lett. **2**, 113–118 (2012)
12. Zhang, J., Majmudar, T.S., Sperl, M., Behringer, R.P.: Jamming for a 2D granular material. Soft Matter. **6**, 2982–2991 (2010)
13. Wood, D.M., Lesniewska, D.: Stresses in granular materials. Granul. Matter. **13**, 395–415 (2011)
14. Lesniewska, D., Wood, D.M.: Photoelastic and photographic study of a granular material. Geotechnique. **61**, 605–611 (2011)
15. Zhang, L., Cai, S., Hu, Z., Zhang, J.: A comparison between bridges and force-chains in photoelastic disk packing. Soft Matter. **10**, 109–114 (2014)
16. Mirbagheri, S.A., Ceniceros, E., Jabbarzadeh, M., McCormick, Z., Fu, H.C.: Sensitively photoelastic biocompatible gelatin spheres for investigation of locomotion in granular media. Exp. Mech. **55**, 427–438 (2015)
17. Mott, P.H., Giller, C.B., Fragiadakis, D., Rosenberg, D.A., Roland, C.M.: Deformation of polyurea: where does the energy go? Polymer. **105**, 227–233 (2016)
18. Lachhab, A., Robin, E., Le Cam, J.B., Mortier, F., Tirel, Y., Canevet, E.: Thermomechanical analysis of polymeric foams subjected to cyclic loading: anelasticity, self-heating and strain-induced crystallization. Polymer. **126**, 19–28 (2017)
19. Chaiamarit, C., Balandraud, X., Preechawuttipong, I., Grédiac, M.: Stress network analysis of 2D non-cohesive polydisperse granular materials using infrared thermography. Exp. Mech. **39**, 761–769 (2015)
20. Jongchansitto, P., Balandraud, X., Grédiac, M., Beitone, C., Preechawuttipong, I.: Using infrared thermography to study hydrostatic stress networks in granular materials. Soft Matter. **10**, 8603–8607 (2014)
21. Jongchansitto, P., Preechawuttipong, I., Balandraud, X., Grédiac, M.: Numerical investigation of the influence of particle size and particle number ratios on texture and force transmission in binary granular composites. Powder Technol. **308**, 324–333 (2017)
22. Halphen, B., Nguyen, Q.S.: Sur les matériaux standard généralisés. Journal de Mécanique. **14**, 39–63 (1975)
23. Samaca Martinez, J.R., Le Cam, J.B., Balandraud, X., Toussaint, E., Caillard, J.: Mechanisms of deformation in crystallizable natural rubber. Part 1: thermal characterization. Polymer. **54**, 2717–2726 (2013)
24. Le Cam, J.B., Samaca Martinez, J.R., Balandraud, X., Toussaint, E., Caillard, J.: Thermomechanical analysis of the singular behavior of rubber: entropic elasticity, reinforcement by fillers, strain-induced crystallization and the Mullins effect. Exp. Mech. **55**, 771–782 (2015)
25. Jongchansitto, P., Balandraud, X., Preechawuttipong, I., Le Cam, J.B., Garnier, P.: Thermoelastic couplings and interparticle friction evidenced by infrared thermography in granular materials. Exp. Mech. (2018, submitted)

Chapter 3
Inverse Identification of the Loading Applied by a Tire on a Landing Gear Wheel

Kévin Cosseron, Renaud Gras, David Mellé, Jean-Frédéric Diebold, François Hild, and Stéphane Roux

Abstract This study aims at identifying the loading applied by a tire on a landing gear wheel for an inflation case. A full scale test instrumented via stereo-DIC (Digital Image Correlation) and strain gages is performed. A 3D finite element model of the wheel is developed and a parameterization of the tire-rim loading is proposed based on model reduction techniques. This parameterization is further used for an inverse identification of the loading parameters. This approach leads to a simpler and more robust problem that can easily be extended to more complex service loadings.

Keywords Stereocorrelation · Inverse identification · Model order reduction · Wheel · Tire

3.1 Introduction

The predictive character of numerical simulations run in order to design a structure critically depends on the accuracy with which the boundary conditions prescribed on the structure are introduced. For landing gear manufacturers, the loadings applied to the wheel by the various tires that can be mounted during ground maneuvering play an important role. However, they are complex and hardly known. The only known quantity is the resultant force between the ground and the tire. Thus, landing gear manufacturers have to make assumptions on the deformation behavior of the tire to design the wheel. The knowledge of the tire-rim interface loadings would hence provide an invaluable help to shorten the design process.

Various attempts have been performed to experimentally investigate the tire-rim interface loadings. Most studies [1, 2] used intrusive instrumentation (strain gages or films) to directly measure contact pressure distribution at the tire-rim interface. However, all of these instrumentations are fragile because of the shear forces encountered in the contact area during tire-wheel assembly or loading and can modify the tire-rim interface behavior. Other investigations [2, 3] moved toward measurements outside the tire-rim contact area (with holographic interferometry or strain gages) to back-calculate the tire-rim loadings combining the experimental results and a finite element or analytical model. Some authors dealt with the modeling of the tire-rim assembly to predict the bead area contact loadings [4, 5]. However, these numerical simulations are highly dependent on the material properties used for the tire model and almost always consider a completely rigid wheel.

The present work is dedicated to the inverse identification of the loading applied to the wheel by a tire for an inflation case via the measurement of the wheel deformation. The aim is to allow the wheel manufacturer to better understand and

K. Cosseron (✉)
LMT, ENS Paris-Saclay, CNRS, Université Paris-Saclay, Cachan, France

Safran Group, Safran Landing Systems, Vélizy-Villacoublay, France
e-mail: cosseron@lmt.ens-cachan.fr

R. Gras
LMT, ENS Paris-Saclay, CNRS, Université Paris-Saclay, Cachan, France

EikoSim, Versailles, France

D. Mellé · J.-F. Diebold
Safran Group, Safran Landing Systems, Vélizy-Villacoublay, France

F. Hild · S. Roux
Laboratoire de Mécanique et Technologie (LMT), ENS Paris-Saclay, Cachan, France

A. Baldi et al. (eds.), *Residual Stress, Thermomechanics & Infrared Imaging, Hybrid Techniques and Inverse Problems, Volume 7*, Conference Proceedings of the Society for Experimental Mechanics Series, https://doi.org/10.1007/978-3-319-95074-7_3

model the interaction between the tire and the wheel, without having to make assumptions on the deformation behavior of the tire. First, the experimental set-up and stereo-DIC measurements are described. Then, a parameterization of the tire-rim loading is proposed using model order reduction techniques. Finally, an identification of the loading is performed based on the comparison of full displacement fields obtained through stereo-DIC on a real test and those computed with a finite element model.

3.2 Experimental Set-Up and Stereo-DIC Measurement

The assembly made of a wheel and a tire is mounted on a testing machine consisting of a moving part and a table to apply vertical and transverse loads respectively. Several quasi-static compression and transverse loadings can be applied with this machine but the loading of the rim due to the inflation of the tire is only studied in the following. A schematic view of this set-up is shown in Fig. 3.1. The contact between the tire and the rim takes place in the inaccessible inner part of the rim. Consequently, it was chosen to observe its outer part (i.e., opposite to the tire contact) in order to be nonintrusive. The test is instrumented with stereo-DIC [6], which gives access to rich kinematic data (i.e. fields), and strain gages. However, the test conditions are challenging for the correlation technique due to the complexity and large scale of the structure to observe (i.e., several meters), and its poor accessibility. A symmetric behavior along the YZ plane is assumed. Ten cameras are positioned around the wheel to simultaneously monitor half of the rim on both sides (inboard and outboard) during the loading. An example of image taken by a camera is shown in Fig. 3.2. The second half of the rim is instrumented with strain gages.

Displacement fields are measured by performing global stereocorrelation analyses on the series of images taken by each camera during the test. A new method of stereocorrelation, compared to the results presented in Ref. [7], is used herein. This method [8] is based on the a priori knowledge of the observed object and uses the object of interest as the calibration target, provided the observed surface has a known definition (i.e., its CAD model). It allows the measured 3D shape and 3D displacement fields to be expressed in the CAD space, thus in the simulation space, and to limit reprojection errors.

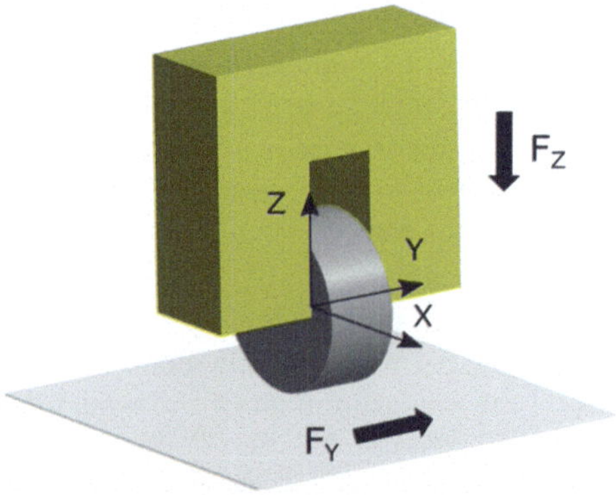

Fig. 3.1 Schematic view of the testing machine

Fig. 3.2 Picture shot by one of the camera

Fig. 3.3 Numerical model

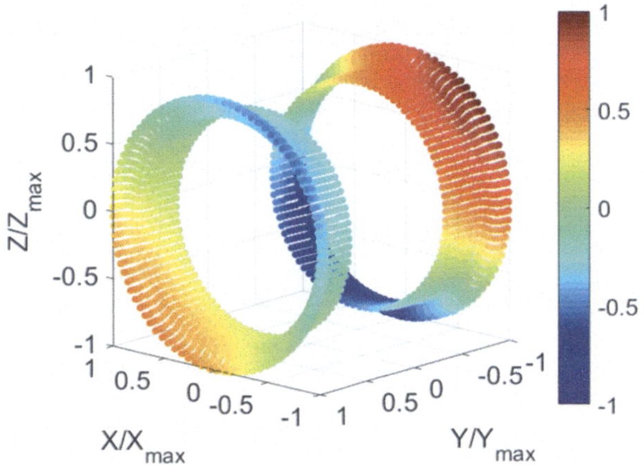

Fig. 3.4 Example of a static eigen mode

3.3 Numerical Model and Loading Parameterization

A 3D finite element model of the wheel (Fig. 3.3) is developed with ANSYS® Workbench™. The tire-rim loading parameterization is performed employing model order reduction techniques. The global stiffness matrix of the structure is reduced to the degrees of freedom of the tire-rim interface using Guyan static condensation [9], and properties of structures with cyclic symmetry [10]. Then, the reduced stiffness matrix is diagonalized to obtain the static eigen modes of the wheel at the tire-rim interface (Fig. 3.4). Thus, any load applied on the rim by a tire is expressed as a linear combination of these eigen modes. A criterion based on the conservation of the elastic strain energy [11] allows the number of the retained modes to be significantly reduced.

3.4 Inverse Identification

An identification of the loading is performed based on the comparison of full-field displacements obtained via stereocorrelation on the real test and those computed with the finite element model. A Finite Element Model Updating algorithm is used. The principle consists in iteratively updating the parameters introduced in the finite element simulations (i.e., static eigen mode amplitudes) in order to minimize a weighted cost function measuring the gap between measured displacements and computed ones. After convergence, the correlation between test and simulation is significantly improved both on displacements in the measurement area and strains at gage locations. The reduced basis composed of the main static eigen modes reveals more compact and robust than other parameterizations. This approach is generic and offers a convenient and efficient framework for arbitrary loadings.

References

1. Walter, J.D., Kiminecz, R.K.: Bead contact pressure measurements at the tire-rim interface. SAE Technical Paper, 750458, 1975
2. Sherwood, J.A., Fussell, B.K., Edwards, W.R., Gross, T.S., Watt, D.W.: Study of the pressure distribution on an aircraft tire-wheel interface. J. Aircr. **32**(5), 921–928 (1995)
3. Kandarpa, S., Spencer, B.F., Kirkner, D.J., Champion, M.: Determination of tire-wheel interface loads for aircraft wheels. J. Aircr. **31**(2), 433–440 (1994)
4. Chen, C.H.S.: Prediction of bead area contact load at the tire-wheel interface using NASTRAN. Tenth NASTRAN® Users' Colloquium, pp. 133–142, 1982
5. Tseng, N.T., Pelle, R.G., Chang, J.P.: Numerical simulation for the tire-rim interface for a radial passenger tire. SAE Technical Paper, 890641, 1989
6. Sutton, M.A., Orteu, J.-J., Schreier, H.: Image Correlation for Shape, Motion and Deformation Measurements: Basic Concepts, Theory and Applications. Springer, US (2009)
7. Gras, R., Hild, F., Mellé, D.: Loading identification on a tire/rim contact for an inflation pressure. In: Proceedings of the 10th International Conference on Advances in Experimental Mechanics (2015)
8. Dufour, J.-E., Hild, F., Roux, S.: Shape, displacement and mechanical properties from isogeometric multiview stereocorrelation. J. Strain Anal. Eng. Des. **50**(7), 470–487 (2015)
9. Guyan, R.J.: Reduction of stiffness and mass matrices. AIAA J. **3**(2), 380–380 (1965)
10. Thomas, D.L.: Dynamics of rotationally periodic structures. Int. J. Numer. Methods Eng. **14**, 81–102 (1979)
11. Sobieszczanski-Sobieski, J., Hajela, P.: Accuracy of an approximate static structural analysis technique based on stiffness matrix eigenmodes. In: Proceedings of the 20th Conference on Structures, Structural Dynamics, and Materials, pp. 127–136. AIAA (1979)

Chapter 4
Fatigue Properties Assessment of API 5L Gr. B Pipeline Steel Using Infrared Thermography

V. E. L. Paiva, R. D. Vieira, and J. L. F. Freire

Abstract Simple and fast procedures to measure the fatigue limit using infrared thermography were proposed in the literature in the last two decades ago. In general, they consider fatigue damage as an energy dissipation process that is accompanied by some temperature variation ΔT. Those procedures significantly reduce fatigue testing costs by decreasing the quantity of required specimens and by shortening the testing time. The aim of the present work is to evaluate the fatigue limit, the stress amplitude vs. number of cycles fatigue (*S-N*) curve, and the influence of mean stress on the fatigue strength. Pipeline steel API 5L Gr. B, very common in the pipeline industry, was the test material. Uniaxial tensile specimens were tested under quasi-static monotonic load or under cyclic loads for different minimum to maximum load ratios. During each test, the surface temperature of the specimens was recorded in real time by a microbolometer thermocamera.

Keywords Infrared · Thermography · Fatigue · Damage · Pipeline material

4.1 Introduction

Two simple procedures to measure the fatigue limit using infrared IR thermography were proposed by Risitano and co-workers [1–4]. In [1–3], uniaxial tensile specimens are subjected to cyclic axial loading. In [4], the initiation of plasticity behavior is captured by the temperature measurements performed during a monotonic quasi-static tensile test to determine the material fatigue strength for full alternated cyclic load ratio ($R = \text{min load/max load} = -1$). The so called Ristano's rapid fatigue assessment procedure encounter other well developed similar approaches, as for example, by Luong [5], Krapez and Pacou [6] and de Finis et al. [7]. The cyclic and quasi-static rapid assessment of the fatigue limit consider the fatigue damage as an energy dissipation process that is accompanied by some temperature variation ΔT. In the general cyclic procedure a test specimen is subjected to a number of blocks of constant stress amplitude loading cycles, which are gradually increased until they eventually cause failure. The quasi-static method is based on the first deviation of the slop of the temperature curve that occurs in the uniaxial tension test. The uniaxial stress value related to the slope deviation point is the full alternate fatigue strength, valid for $R = -1$. Both methods significantly reduce the traditional fatigue testing costs by decreasing the quantity of required specimens and by shortening the testing time. The aim of the present work is to evaluate the fatigue limit, the stress amplitude vs. number of cycles fatigue (*S-N*) curve, and the influence of mean stress on the fatigue strength. Pipeline steel API 5L Gr. B, very common in the pipeline industry, was the test material.

4.2 Material and Experimental Procedure

Uniaxial specimens were tested in a 100 kN INSTRON servo-hydraulic machine under quasi-static monotonic load or cyclic axial load. The stroke rate applied in the static tests was 1 mm/min. Based on the cyclic experimental methodology proposed, the cyclic tests employed stepped loading cycles, sequentially applied to each specimen at a constant amplitude force-control, with 5–15 Hz frequency and varying load ratios *R*. During each quasi-static or cyclic test, the surface temperature of the specimen was recorded in real time by a microbolometer thermocamera FLIR A655sc (640 × 480 uncooled micro-bolometers, 50 Hz acquisition rate, 17 μm spatial resolution, 30 mK sensitivity). The temperature data was acquired and analyzed using the ResearchIR software from FLIR.

V. E. L. Paiva · R. D. Vieira · J. L. F. Freire (✉)
Pontifical Catholic University of Rio de Janeiro, PUC-Rio, Rio de Janeiro, Gávea, RJ, Brazil
e-mail: jlfreire@puc-rio.br

A. Baldi et al. (eds.), *Residual Stress, Thermomechanics & Infrared Imaging, Hybrid Techniques and Inverse Problems, Volume 7*, Conference Proceedings of the Society for Experimental Mechanics Series, https://doi.org/10.1007/978-3-319-95074-7_4

Table 4.1 Material properties

Chemical composition	0.15C, 0.83Mn, 0.19Si, 0.02P, 0.003S
Yield strength S_y (average of 3 specimens)	316 MPa (0.5% total engineering strain)
Ultimate strength S_u (average of 3 specimens)	420 MPa

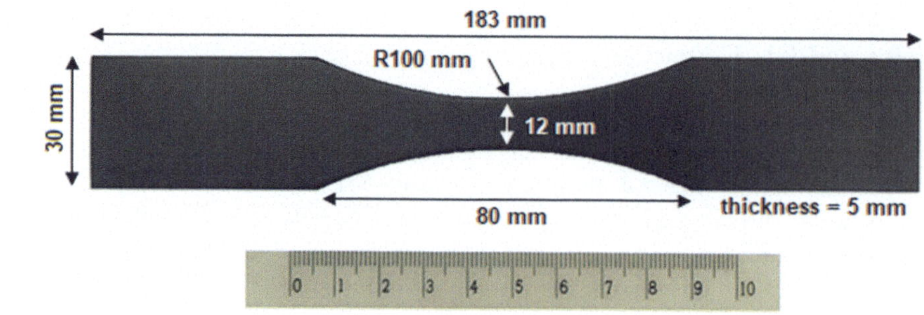

Fig. 4.1 Specimen's geometry

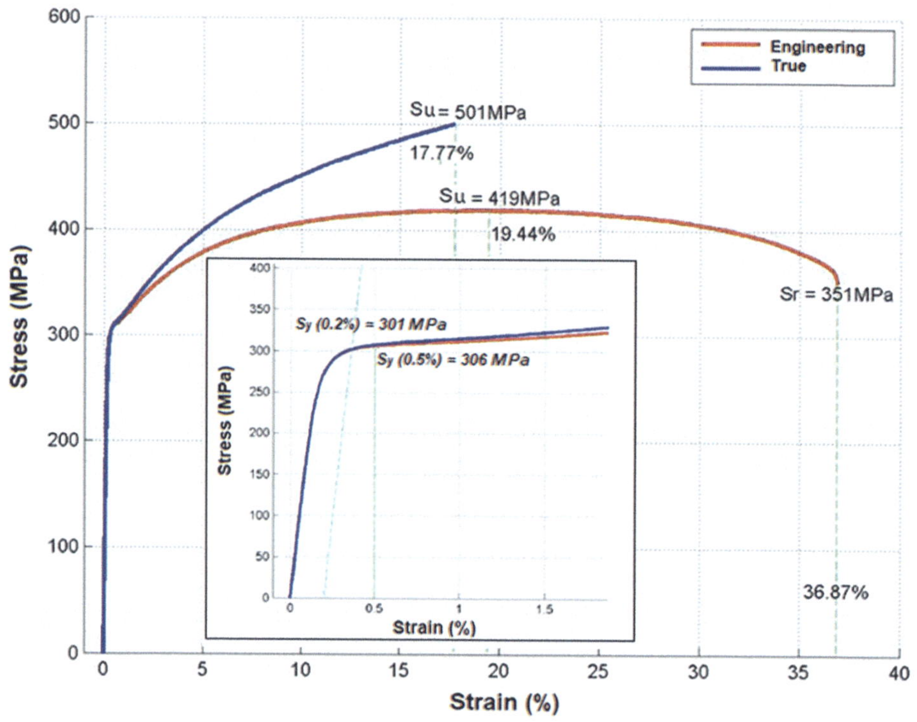

Fig. 4.2 Tensile stress-strain curve of one of the 3 specimens of API 5L Gr. B pipe

Relevant material data are given in Table 4.1. The uniaxial test specimens were machined from longitudinal strips cut from a thin walled API 5L Gr. B pipe having a nominal outside diameter D = 324 mm and thickness t = 6.35 mm. Figure 4.1 gives the final nominal specimen's dimensions following ASTM E 466. All uniaxial specimens were painted with a thin layer of opaque black paint to increase emissivity. The uniaxial tensile stress-strain curve for the pipe material is given in Fig. 4.2.

4.3 IR Thermography

According to the methodology proposed in [1–3], when the specimen is loaded below the fatigue limit, its temperature varies very slightly, but for alternated stress amplitudes ($\Delta\sigma_i = \sigma_{ai}$) above the fatigue limit the temperature variations are significant. The temperature variation according to a given reference (ΔT) may be classified along the test inside three

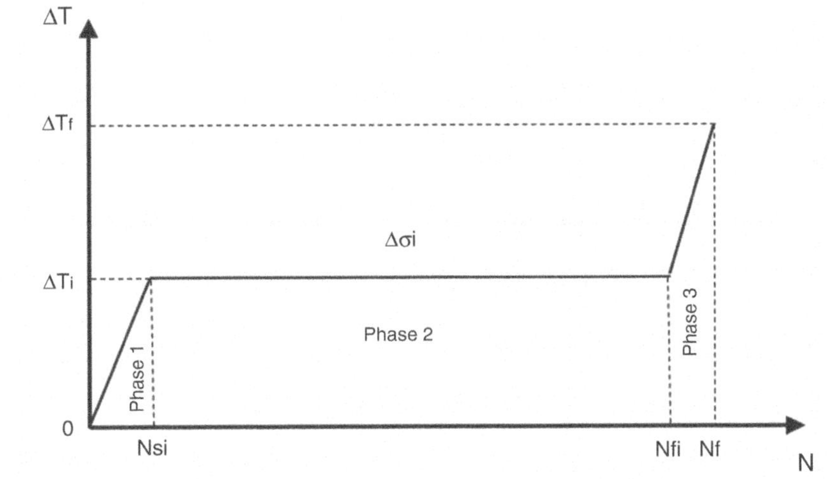

Fig. 4.3 Phases of the thermal behavior of ΔT vs. N curve for a hot spot in the specimen surface during typical fatigue tests. (Adapted from Ref. [2])

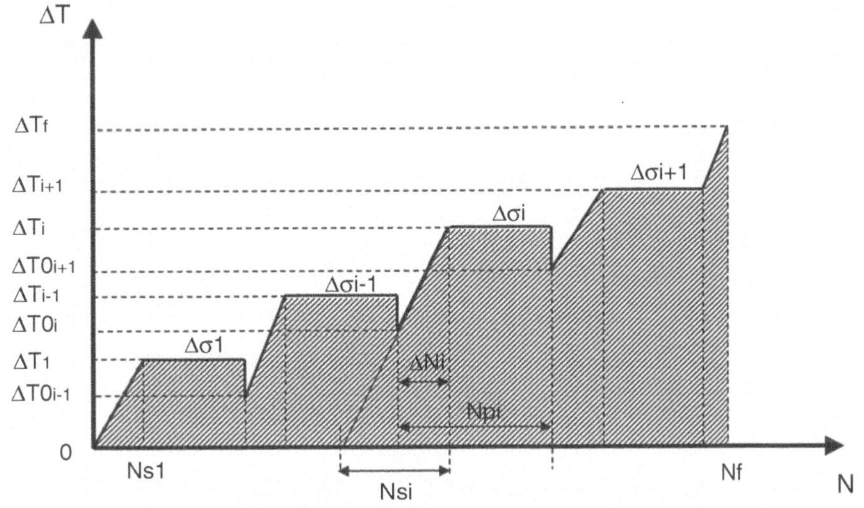

Fig. 4.4 Stepped loading procedure for the determination of the fatigue limit and the integral ϕ

phases. Figure 4.3 presents the temperature variation with the number of cycles N for the test where the stress amplitude $\Delta\sigma_i$ was made constant. They increase during the first part of the test (phase 1), then remain stable for a while (phase 2), and finally rapidly increase prior to failure at a life of N_f cycles (phase 3), as shown in Fig. 4.3.

Through the analysis of the temperature response curve under incremental cyclic step loading, see Figs. 4.4 and 4.5a, b, the fatigue limit can be evaluated. Each stress amplitude level corresponds to one stabilization temperature. Two linear regression lines are used to approximate the thermal data and to determine graphically the fatigue limit, as depicted in Fig. 4.5b. The first line contains the data where the applied stress is below the fatigue limit, and the second line fits the data located above the limit. The fatigue limit is determined by the intersection of these two lines, as shown in Fig. 4.5b.

Using the same line of analysis, Fargione et al. [2] proposed a method to determine the fatigue-life curve *S-N* of the material using IR thermography. This method is much faster than the traditional one and needs theoretically only one specimen. In practice, the recommended minimum number of specimens is three. For small temperature variations (under 100 K, implicating small loading frequencies), the heat transferred from the specimen to the environment can be considered proportional to the temperature difference ΔT. A parameter ϕ, which is the integral of the ΔT vs. N curve in Fig. 4.3, Eq. (4.1),

$$\phi = \int_0^{N_f} \Delta T dN \tag{4.1}$$

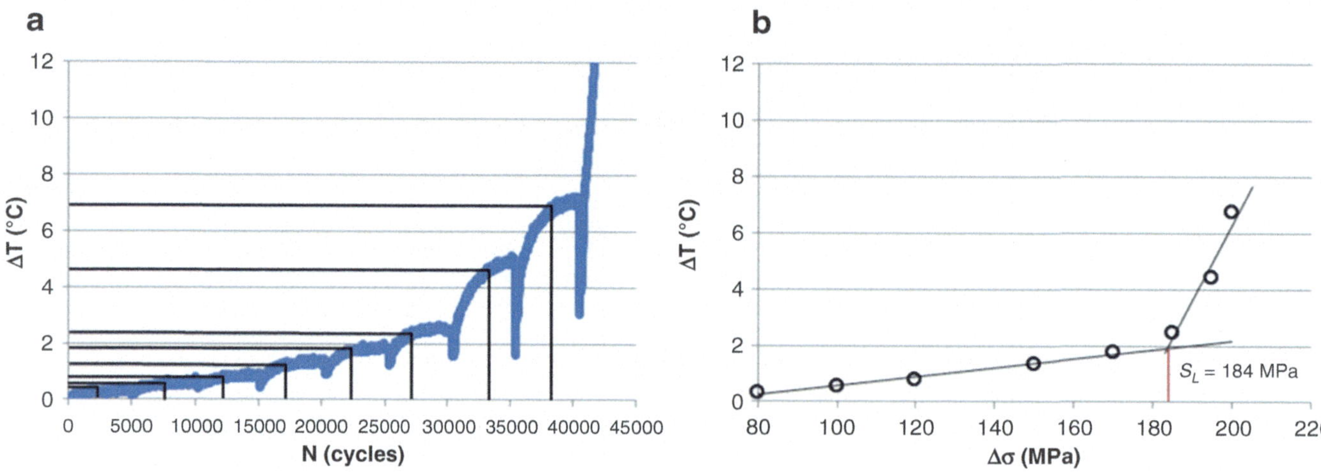

Fig. 4.5 (**a**) ΔT vs. N curve for various stress amplitudes $\Delta\sigma_i$, with the determination of each ΔT_i. (**b**) ΔT_i vs. $\Delta\sigma_i$ curve for various incremental load steps and determination of the fatigue limit (intercept of the two straight lines). Data obtained for specimen 2 of the present paper, $R = 0.025$

is assumed constant for given test conditions, such as the specimen geometry and material, loading frequency and environment. This critical value corresponds to fatigue failure and it is associated to the total area under the curve of Fig. 4.3 (for one constant stress amplitude level), or the total dashed area in Fig. 4.4 (for several stress amplitude levels). The ϕ value can be used together with values of ΔT_i, ΔT_{0i}, ΔN_i measured at stress amplitude levels $\Delta\sigma_i$ for a single specimen, as depicted in Fig. 4.3 and modeled by Eqs. (4.2) and (4.3), to yield the whole fatigue *S-N* curve. Each data point of the *S-N* curve is formed by the pair (N_{fi}, $\Delta\sigma_i$) where N_{fi} is calculated from Eq. (4.2). This equation is derived with the help of Fig. 4.4 and the assumption that phase 3, depicted in Fig. 4.3, is small and can be neglected. The value of N_{si} is determined from Eq. 4.3 with the help of Fig. 4.4.

$$\phi = \Delta T_i \left(N_{fi} - \frac{\Delta N_{si}}{2} \right) \tag{4.2}$$

$$N_{si} = \frac{\Delta N_i \Delta T_i}{\Delta T_i - \Delta T_{0i}} \tag{4.3}$$

A partial damage value D_i results from the partial integrated value ϕ_i resulting from the application of each stress amplitude $\Delta\sigma_i$. The value ϕ_i is calculated considering the number of cycles applied at each stress level, N_{pi}. The denominator of Eq. (4.4) gives the area under the curve ΔT vs. N considering the interval N_{pi}. The partial damage value of Eq. (4.4) assumes that the Miner's linear damage accumulation rule is valid.

$$D_i = \frac{\phi_i}{\phi} = \frac{\left(\frac{\Delta T_i + \Delta T_{0i}}{2}\right) \Delta N_i + \Delta T_i \left(N_{fi} - \Delta N_{si}\right)}{\sum_1^f \phi_f} \tag{4.4}$$

Based on the experimental cyclic methodology [1–3], next section presents results for four specimens that were tested under a step loading procedure, sequentially applied to each specimen at a constant amplitude under force-control, with stress ratios $R = \sigma_{min}/\sigma_{max}$ equal to $R = 0.025$ and 0.5. At each load step, the load amplitude was maintained fixed during blocks of $5 \cdot 10^3$ cycles, a value that was high enough for the temperature to achieve a stable thermal behavior, as determined by preliminary tests. Afterwards, the load was increased until the specimen failed. One specimen was tested based on the quasi-static methodology [4]. Results from both cyclic and quasi-static analyses are presented in the next section.

4.4 Results and Discussion

Results of tests performed to characterize the uniaxial material behavior are given in Figs. 4.2, 4.5, 4.6, and 4.7 and reported in Tables 4.1, and 4.2. Figure 4.2 gives the uniaxial material stress-strain curve for a monotonic test. The detailed plot inside the larger plot shows the method used to determine the 0.5% yield strength for a longitudinal test specimen.

Figure 4.5 depicts the determination of the fatigue limit by applying the Risitano's rapid method [1–3]. The temperature variation with number of cycles and with applied stress amplitude are presented respectively in Fig. 4.5a, b. These plots show that the temperature undergoes a significant increase when it is submitted to elevated stress levels, above the fatigue limit. Figure 4.5a shows that temperature stabilizes around 2300 cycles for each stress level, completing the transition from phase 1 to phase 2, as showed in Fig. 4.3. Figure 4.5a shows that, after reaching temperature stabilization, the cyclic loading is kept for approximately 2700 cycles until the test stress level is changed. This way the specimen is consecutively loaded until it finally reaches its failure (phase 3 in Fig. 4.3).

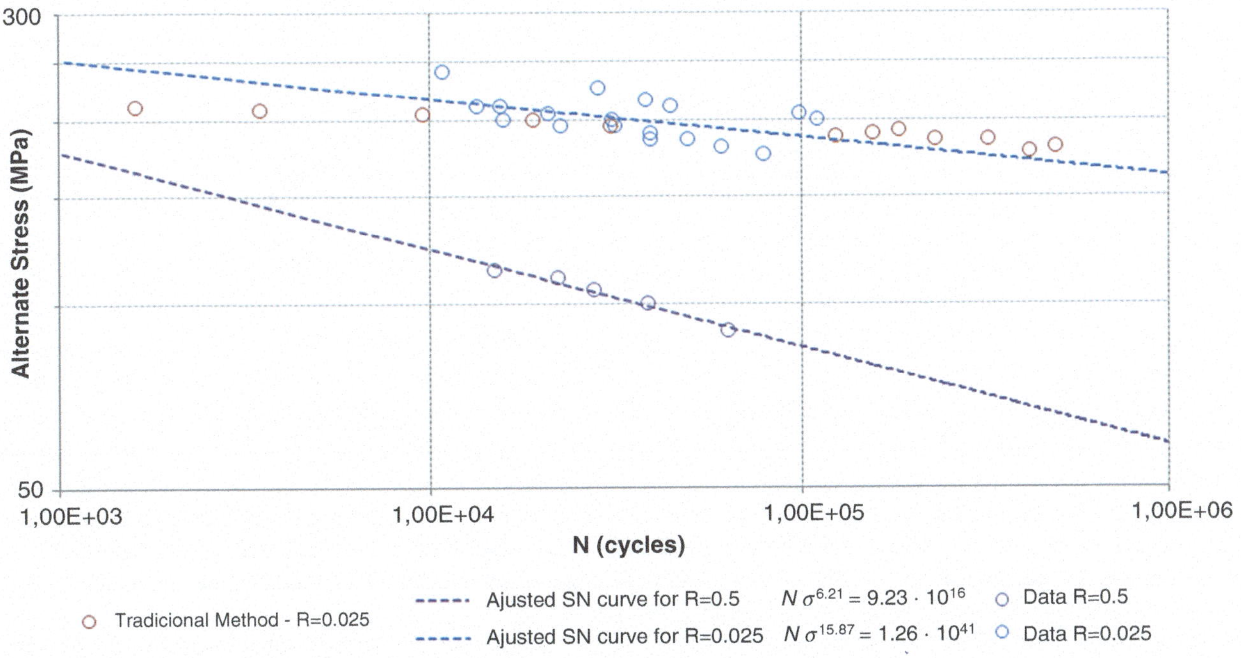

Fig. 4.6 Fatigue *S-N* curve for stress ratios $R = 0.025$ and $R = 0.5$

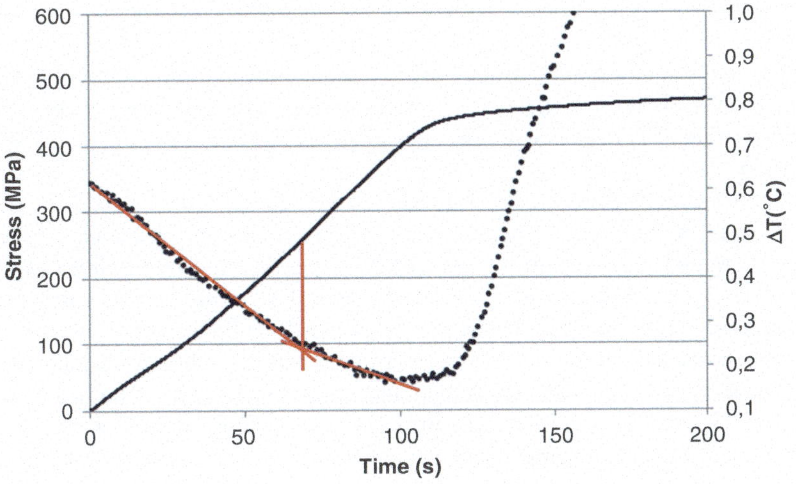

Fig. 4.7 Fatigue limit ($R = -1$) for steel API 5 L Gr. B using the quasi-static method proposed in [4]

Table 4.2 Measured results for each specimen

Specimen	Fatigue limit S_L (MPa)	R	Frequency (Hz)	Test type	ϕ (°C·cycles)	D (from Eq. (4.4))
1	169	0.025	15	Cyclic	1.53×10^5	0.98
2	184	0.025	15		9.2×10^4	0.97
3	86	0.5	15		3.0×10^4	0.96
6	196	0.025	5		5.2×10^5	0.95
5	253	−1	1 mm/min stroke	Quasi-static	–	–

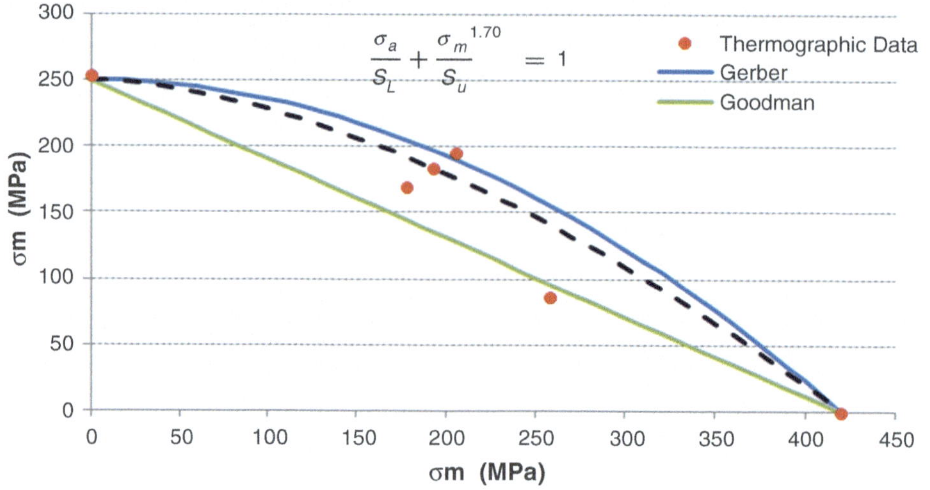

Fig. 4.8 Plot for alternate stress vs. mean stress for Gerber and Goodman lines and for thermographic data

Figure 4.5b shows the stabilized temperatures, ΔT_i, obtained from Fig. 4.5a, plotted against their corresponding amplitude stress levels, $\Delta\sigma_i$. According to Ristanos' method, the fatigue limit S_L can be determined graphically by the intersection of two regression lines based on the data below and above the fatigue limit, as illustrated in Fig. 4.5b. When the specimen is subjected to a stress amplitude above the fatigue limit, its temperature increases significantly and damage begins to accumulate at one or more critical hotspots until the formation and propagation of a crack that leads to the specimen failure. Figure 4.5b shows the fatigue limit $S_L = 184$ MPa determined for specimen 2 under $R = 0.025$. Values for other specimens tested under ratios R equal to 0.025 and 0.5 are given in Table 4.2.

Having determined the values of ϕ as presented in Eq. (4.2), the *S-N* fatigue curve can be determined using the process outlined in the previous section. The *S-N* fatigue curves for each stress ratio R calculated in this way are presented in Fig. 4.6. The calculated curves encompass data fitted to the interval 10^3 to 10^6 cycles. Traditional fatigue tests were also performed using the ratio $R = 0.025$ and are plotted in Fig. 4.6. Table 4.2 presents the fatigue results for all tested specimens using the rapid assessment method.

Figure 4.7 depicts the quasi-static determination of the full alternated uniaxial fatigue limit of the pipe material as proposed by Risitano et al. [4]. The method is based on the first deviation of the slop of the temperature curve that occurs in the uniaxial tension test. This slope deviation point is related with the fully alternate fatigue strength, valid for $R = -1$. This value was found to be $S_{L\_R=-1} = 253$ MPa.

It is interesting to note that the results for $R = -1$ (from the static test), $R = 0.025$ and $R = 0.5$ (from the cyclic tests), together with the material ultimate strength S_u, are compatible with the mean stress influence on the fatigue limit when seen under the light of a plot of alternated vs. mean stress, presented in Fig. 4.8. In this Figure, Goodman and Gerber curves, computed using the values of $S_L = 253$ MPa for $R = -1$ and $S_u = 420$ MPa, are plotted for comparison purpose. Similar results have already been published by Krapez and Pacou [6] for steel X48 and Rego et al. [8] for brass 36,000.

Based on the fatigue limits presented in Table 4.2 for each applied R, data points formed by pairs of alternate stress amplitude values (σ_a) and mean stress (σ_m) were calculated and plotted in Fig. 4.8. The resulting data was adjusted by a parabolic fitting equation presented in Eq. (4.5), the adjusting exponent α being equal to 1.70. This exponent falls between the exponents given by the so called Goodman and Gerber curves, respectively equal to 1 and 2 [9].

$$\frac{\sigma_a}{S_L} + \left(\frac{\sigma_m}{S_u}\right)^{\alpha} = 1 \tag{4.5}$$

4.5 Conclusions

This paper confirms that rapid fatigue damage assessment methods are practical and efficient tools that can provide a reliable and fast way to determine the fatigue behavior of materials using few specimens. More specifically, the paper furnishes fatigue data for the API 5L Gr. B pipeline steel, which is scarce in the literature. Moreover, the influence of the stress ratio R was evaluated for this material and an equation showing the influence of the mean applied stress on the fatigue limit was given, which can be useful for design purpose.

References

1. La Rosa, G., Risitano, A.: Thermographic methodology for rapid determination of the fatigue limit of materials and mechanical components. Int. J. Fatigue. **22**, 65–73 (2000)
2. Fargione, G., Geraci, A., La Rosa, G., Risitano, A.: Rapid determination of the fatigue curve by the thermographic method. Int. J. Fatigue. **24**, 11–19 (2000)
3. Risitano, A., Risitano, G.: Cumulative damage evaluation of steel fracture mechanics. Theor. Appl. Fract. Mech. **54**, 82–90 (2010)
4. Risitano, G., Risitano, A., Clienti, C.: Determination of the fatigue limit by semi static tests. Convegno Nationale, IGF XXI, Cassino, Italia 13–15 Giugo 2011, pp. 322–330
5. Luong, N.P.: Infrared thermography of fatigue in metals. SPIE vol 1682 Thermosense XIV
6. Krapez, J.C., Pacou, D.: Thermographic detection of damage initiation during fatigue tests. SPIE vol 4710 2002 Thermosense XXIV
7. de Finis, R., Palumbo, D., Ancona, F., Galietti, U.: Fatigue limit evaluation of various martensitic stainless with new robust thermographic data analysis. Int. J. Fatigue. **74**, 88–96 (2015)
8. Rego, L.L.L., Castro, J.T.P., Freire, J.L.F., Paiva, V.E.L.: Fatigue characterization of the C36000 copper alloy using the thermographic method. In: 24th ABCM International Congress of Mechanical Engineering, 3–8 Dec, 2017, Curitiba
9. Castro, J.T.P., Meggiolaro, M.A.: Fatigue Design Techniques, Volume 1: High-Cycle Fatigue. CreateSpace Independent Publishing Platform (2016)

Chapter 5
Determination of Constitutive Parameters in Inverse Problem Using Thermoelastic Data

Abdullah A. Alshaya and John M. Considine

Abstract A new inverse problem formulation for identification of constitutive parameters in orthotropic materials from load-induced thermal information is developed using Levenberg-Marquardt Algorithm and Airy stress function. Inverse methods were used to determine the constitutive properties as well as the thermoelastic calibration factors of a loaded perforated graphite/epoxy laminated composite by processing noisy simulated thermoelastic data with an Airy stress function in complex variables. Equilibrium, compatibility, and traction-free condition on the boundary of the circular hole are satisfied using complex-variable formulation, conformal mapping and analytic continuation. The primary advantage of this new formulation is the direct use of load-induced thermal data to determine the constitutive parameters, separate the stresses, i.e., evaluate the individual stress components, including on the edge of the hole, and smooth the measured data, all from a single test. The inverse method algorithm determined the constitutive properties with errors less than 10%.

Keywords Composites · Airy Stress Function · Inverse Problems · Thermoelastic Stress Analysis · Complex Variables

5.1 Introduction

The Airy stress function in complex variables was used extensively in determining stresses from measured data [1–6]. The Airy stress function can be processed with measured data using thermoelasticity (thermoelastic stress analysis, TSA) [1, 2], photoelasticity [3], digital image correlation [4], moiré [5] or strain gages [6]. These hybrid methods do not necessitate knowing the applied loads or external geometry. In addition, the proposed hybrid methods smooth the measured data and determine the individual stresses throughout, including on the edge of the hole. All of the prior applications of the mapping technique evaluated the stresses using the constitutive properties and TSA calibration factors found experimentally from standard tensile tests whereas the present approach only evaluated these parameters using the thermoelastic data.

In general, identification of a material constitutive parameters requires the use of inverse methods (IM). Avril and Pierron [7] reviewed several IM approaches and showed their general equivalency. Alshaya et al. [8] determined the constitutive properties of a symmetrically sided notched graphite/epoxy composite using recorded DIC displacement data and Airy stress function. Inverse method (IM) can be generally described as the iterative adjustments of parameters (constitutive properties) in a numerical model (in this case, an Airy stress function scheme) to minimize the difference between an experimentally measured quantity (thermoelastic data) and the numerically calculated quantity. In 2-D models, the degree of freedom is (number of nodes) $\times$ 2 – (number of constitutive parameters) – 1. For homogeneous, isotropic materials, the number of constitutive properties is two (E, ν); for homogeneous, orthotropic materials, the number of constitutive parameters are four (E_{11}, E_{22}, G_{12}, ν_{12}). For either case, the number of degrees of freedom is large and the problem is solved by minimizing least squares of the chosen cost function. The goal of this work is to evaluate the constitutive properties of a composite plate as well as the thermomechanical calibration factors of TSA using IM and Airy stress function scheme. The primary difference IM technique described here is that the specimen geometry is chosen so that an Airy stress function is known a priori and, therefore, the problem is statically determinant. The authors are unaware of prior utilization of mapping and complex variables to determine the constitutive properties in composites from thermoelastic data as well as the TSA calibration factors.

A. A. Alshaya (✉)
Department of Mechanical Engineering, Kuwait University, Safat, Kuwait
e-mail: alshaya@wisc.edu

J. M. Considine
USDA, Forest Service, Forest Products Laboratory, Madison, WI, USA

A. Baldi et al. (eds.), *Residual Stress, Thermomechanics & Infrared Imaging, Hybrid Techniques and Inverse Problems, Volume 7*, Conference Proceedings of the Society for Experimental Mechanics Series, https://doi.org/10.1007/978-3-319-95074-7_5

5.2 Thermoelastic Stress Analysis

Thermoelastic stress analysis (TSA) is a non-contacting, non-destructive experimental method for determining the full-field stresses in loaded members. By cyclically loading the structure to satisfy adiabatic reversible conditions, the stresses at a location are related to the stress-induced thermal information at that position. Under orthotropy, the thermoelastic system signal, S^*, is proportional to the following change in the linear combination of the normal stresses, σ_1 and σ_2, in the directions of material symmetry

$$S^* = \Delta\left(K_1\sigma_1 + K_2\sigma_2\right) \tag{5.1}$$

where K_1 and K_2 are the orthotropic thermomechanical coefficients and can be determined experimentally.

5.3 Relevant Equations

5.3.1 Basic Equations

For plane problems having rectilinear orthotropy and no body forces, the Airy stress function, $\mathcal{F}$, can be expressed as a summation of two arbitrary analytical functions, $F_1(z_1)$ and $F_2(z_2)$, of the complex variables, z_1 and z_2, as [9]

$$\mathcal{F} = 2\mathrm{Re}\left[F_1\left(z_1\right) + F_2\left(z_2\right)\right] \tag{5.2}$$

such that $z_j = x + \mu_j y$ for $j = 1, 2$ and *Re* denotes the 'real part' of a complex number. The complex material properties μ_1 and μ_2 are two distinct roots of the following characteristic equation associated with the compatibility equation

$$\mu^4 + \left(\frac{E_{11}}{G_{12}} - 2\nu_{12}\right)\mu^2 + \frac{E_{11}}{E_{22}} = 0 \tag{5.3}$$

The 1- and 2-orientations are the directions of orthotropic material symmetry. The stresses in rectangular coordinates (x, y) of the physical z $(=x + iy)$ plane can be expressed in terms of the stress functions. By introducing the new stress functions

$$\Phi\left(z_1\right) = \frac{dF_1\left(z_1\right)}{dz_1}, \quad \text{and} \quad \Psi\left(z_2\right) = \frac{dF_2\left(z_2\right)}{dz_2} \tag{5.4}$$

one can write the stresses as

$$\sigma_{xx} = 2\mathrm{Re}\left[\mu_1^2\Phi'\left(z_1\right) + \mu_2^2\Psi'\left(z_2\right)\right] \tag{5.5}$$

$$\sigma_{yy} = 2\mathrm{Re}\left[\Phi'\left(z_1\right) + \Psi'\left(z_2\right)\right] \tag{5.6}$$

$$\sigma_{xy} = -2\mathrm{Re}\left[\mu_1\Phi'\left(z_1\right) + \mu_2\Psi'\left(z_2\right)\right] \tag{5.7}$$

where primes denote differentiation with respect to the argument. Plane problems of elasticity classically involve determining the stress functions, $\Phi(z_1)$ and $\Psi(z_2)$, throughout a component and subject to the boundary conditions around its entire edge. For a region of a component adjacent to a traction free-edge, $\Phi(z_1)$ and $\Psi(z_2)$ can be related to each other by the conformal mapping and analytic continuation techniques. The stresses can then be expressed in terms of the single stress function, $\Phi(z_1)$. Moreover, $\Phi(z_1)$ will be represented by a truncated power-series expansion whose unknown complex coefficients are determined experimentally. Once $\Phi(z_1)$ and $\Psi(z_2)$ are fully evaluated, the individual stresses are known from Eqs. (5.5) through (5.7). For a significantly large region of interest in a finite structure, it may also be necessary to satisfy other boundary conditions at discrete locations.

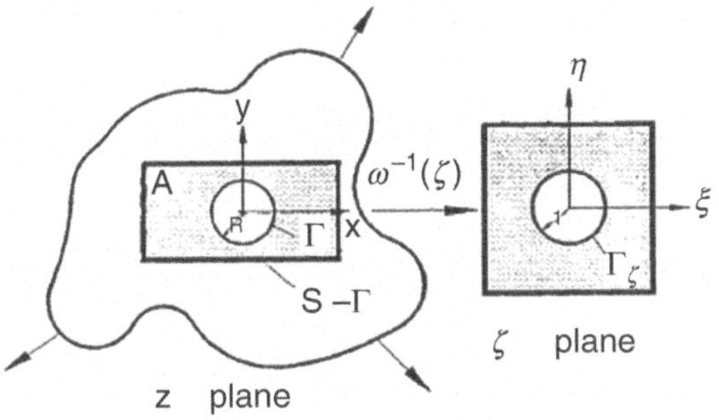

Fig. 5.1 Mapping circular cutout in the physical z-plane into exteior region of a unit circle in ζ-plane

5.3.2 *Conformal Mapping*

Conformal mapping is introduced to simplify the plane problem by mapping the region R_z of a complicated physical $z = x + iy$ plane of a loaded structure into a region R_ζ of a simpler shape in the $\zeta = \xi + i\eta$ plane, the latter being a unit circle if one represents the stress function as a Laurent series, Fig. 5.1 [9–11]. The new coordinate system (and resulting geometry) is usually chosen to aid in solving the equations and the obtained solution from this simplified domain can then be mapped back to the original physical geometry for a valid solution.

Assume that a mapping function of the form $z = \omega(\zeta)$ exists and which maps R_ζ of the simpler plane into R_z of the more complicated physical plane. For orthotropy, auxiliary planes and their induced mapping functions are defined in terms of $\zeta_j = \xi + \mu_j\eta$, therefore $z_j = \omega_j(\zeta_j)$, for $j = 1, 2$. The induced conformal mapping functions are one-to-one and invertible. The stress functions $\Phi(z_1)$ and $\Psi(z_2)$ can be expressed as the following analytic functions of ζ_1 and ζ_2.

$$\Phi(z_1) = \Phi[\omega_1(\zeta_1)] \equiv \Phi(\zeta_1) \qquad \text{and} \qquad \Psi(z_2) = \Psi[\omega_2(\zeta_2)] \equiv \Psi(\zeta_2) \tag{5.8}$$

The derivatives of the stress functions with respect to their argument are

$$\Phi'(z_1) = \Phi'(\zeta_1)\frac{d\zeta_1}{dz_1} = \frac{\Phi'(\zeta_1)}{\omega_1'(\zeta_1)} \qquad \text{and} \qquad \Psi'(z_2) = \frac{\Psi'(\zeta_2)}{\omega_2'(\zeta_2)} \tag{5.9}$$

The analyticity of the mapping functions satisfies the equilibrium and compatibility throughout region R_z of the physical plane.

5.3.3 *Traction–Free Boundaries*

Using the concept of analytic continuation, the stress functions for a region R_ζ adjacent to a traction-free boundary of the unit circle of an orthotropic material are related by [12]

$$\Psi(\zeta_2) = B\overline{\Phi\left(1/\overline{\zeta}_2\right)} + C\Phi(\zeta_2) \tag{5.10}$$

where constants B and C are the following complex material properties

$$B = \frac{\overline{\mu}_2 - \overline{\mu}_1}{\mu_2 - \overline{\mu}_2}, \quad C = \frac{\overline{\mu}_2 - \mu_1}{\mu_2 - \overline{\mu}_2} \tag{5.11}$$

Equation (5.10) enable the elastic state of the structure to be expressed in terms of a single stress function, $\Phi(\zeta_1)$, the latter which can be represented by a Laurent series expansion. Equation (5.10) assumes ability to map the physical boundary of interest into the unit circle in the mapped plane. Reference [13] contains a more thorough derivation of Eq. (5.10).

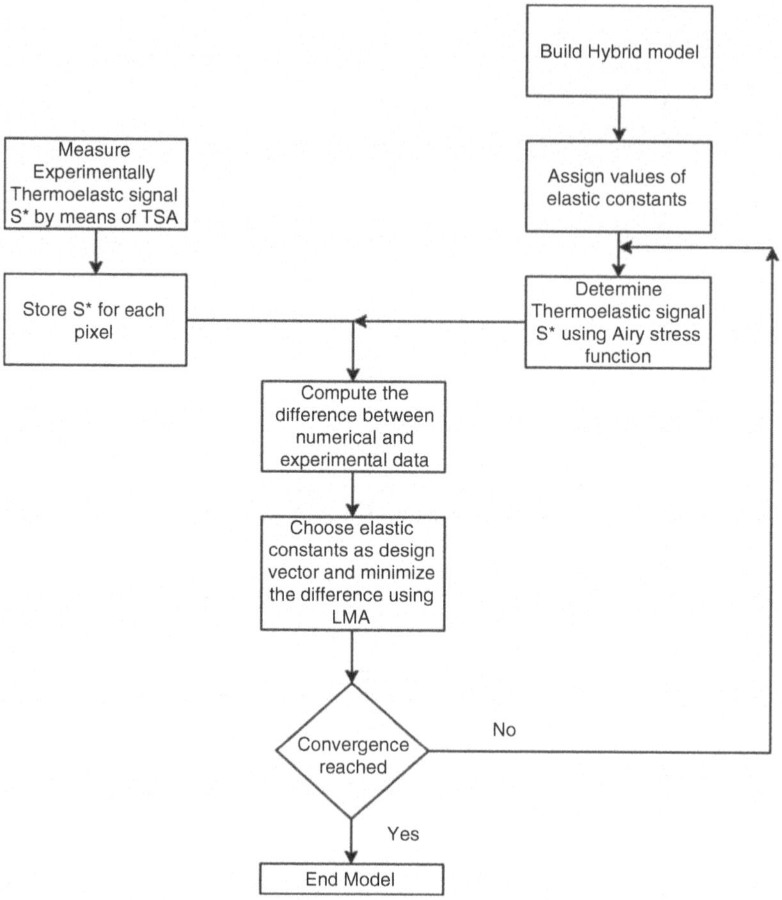

Fig. 5.2 Flow chart of the inverse problem procedure for orthotropic material characterization using Airy stress function scheme

5.3.4 Mapping Formulation

The objective here is to apply the approach to a region R_z adjacent to a traction-free boundary of a physical member provided an appropriate mapping function is available to map the region R_ζ into region R_z, where Γ_ζ, the exterior of a unit circle in the ζ-plane, goes to the physical traction-free boundary, Γ. For a region adjacent the circular notch of radius R, the following function

$$z_j = \omega_j\left(\zeta_j\right) = \frac{R}{2}\left[\left(1 - i\mu_j\right)\zeta_j + \frac{1 + i\mu_j}{\zeta_j}\right], \quad j = 1, 2 \tag{5.12}$$

maps the region of the exterior of a unit circle, R_ζ, of the ζ-plane into the region R_z of the z-physical plane, Fig. 5.2. The inverse of the induced mapping function is

$$\zeta_j = \omega_j^{-1}\left(z_j\right) = \frac{z_j \pm \sqrt{z_j^2 - R^2\left(1 + \mu_j^2\right)}}{R\left(1 - i\mu_j\right)}, \quad j = 1, 2 \tag{5.13}$$

The branch of the square root in Eq. (5.13) is chosen such that $|\zeta_j| \geq 1$ for $j = 1, 2$.

5.3.5 Mapping Collocation

The single stress function can be expresses as the following finite Laurent series

$$\Phi\left(\zeta_1\right) = \sum_{\substack{j=-N \\ j \neq 0}}^{N} A_j \zeta_1^j \tag{5.14}$$

where $A_j = a_j + ib_j$ are the unknown complex coefficients (a_j and b_j are both real numbers). The $j = 0$ term contributes to rigid-body motion and can be omitted. Substituting Eq. (5.14) into (5.10) yields

$$\Psi\left(\zeta_2\right) = \sum_{\substack{j=-N \\ j \neq 0}}^{N} \left(\overline{A}_j B \zeta_2^{-j} + A_j C \zeta_2^j\right) \tag{5.15}$$

where $\overline{A}_j$ is the complex conjugate of A_j. For a finite, simply connected region R_ζ, $\Phi(\zeta_1)$ is a single-valued analytic function. Orthotropic composite complex parameters are purely imaginary when the directions of material symmetry are parallel and perpendicular to the applied load and require that only odd terms be retained in the Laurent expansions. From Eqs. (5.5) through (5.8), the stresses can be written as

$$\sigma_{xx} = 2 \sum_{\substack{j=-N,-N+2,\ldots \\ j \neq 0}}^{N} \operatorname{Re}\left\{ j \left[\frac{\mu_1^2 \zeta_1{}^{j-1}}{\omega_1'\left(\zeta_1\right)} + \frac{C \mu_2^2 \zeta_2{}^{j-1}}{\omega_2'\left(\zeta_2\right)} \right] A_j - j \mu_2^2 B \left[\frac{\zeta_2{}^{-j-1}}{\omega_2'\left(\zeta_2\right)} \right] \overline{A}_j \right\} \tag{5.16}$$

$$\sigma_{yy} = 2 \sum_{\substack{j=-N,-N+2,\ldots \\ j \neq 0}}^{N} \operatorname{Re}\left\{ j \left[\frac{\zeta_1{}^{j-1}}{\omega_1'\left(\zeta_1\right)} + \frac{C \zeta_2{}^{j-1}}{\omega_2'\left(\zeta_2\right)} \right] A_j - j B \left[\frac{\zeta_2{}^{-j-1}}{\omega_2'\left(\zeta_2\right)} \right] \overline{A}_j \right\} \tag{5.17}$$

$$\sigma_{xy} = -2 \sum_{\substack{j=-N,-N+2,\ldots \\ j \neq 0}}^{N} \operatorname{Re}\left\{ j \left[\frac{\mu_1 \zeta_1{}^{j-1}}{\omega_1'\left(\zeta_1\right)} + \frac{C \mu_2 \zeta_2{}^{j-1}}{\omega_2'\left(\zeta_2\right)} \right] A_j - j \mu_2 B \left[\frac{\zeta_2{}^{-j-1}}{\omega_2'\left(\zeta_2\right)} \right] \overline{A}_j \right\} \tag{5.18}$$

The only unknowns in these expressions for the stresses are the complex coefficients, A_j. The latter can be determined from thermoelastic data. Choosing the x-axis parallel to the stiffest orientation of the composite, i.e., 1-direction of an orthotropic composite material, the TSA signal S^*, Eq. (5.1), can be expressed as

$$S^* = \mathrm{K}_1 \sigma_{xx} + K_2 \sigma_{yy} = 2 \sum_{\substack{j=-N,-N+2,\ldots \\ j \neq 0}}^{N} \operatorname{Re}\left\{ \left[\frac{j\left(K_1 \mu_1^2 + K_2\right)}{\omega_1'\left(\zeta_1\right)} \zeta_1{}^{j-1} + \frac{j\left(K_1 \mu_2^2 + K_2\right) C}{\omega_2'\left(\zeta_2\right)} \zeta_2{}^{j-1} \right] A_j - \left[\frac{j\left(K_1 \mu_2^2 + K_2\right) B}{\omega_2'\left(\zeta_2\right)} \zeta_2{}^{-j-1} \right] \overline{A}_j \right\} \tag{5.19}$$

The only unknowns in these expressions for the stresses are the complex coefficients $A_j = a_j + ib_j$, the other quantities involve geometry (location) or material properties. Because the summation in Eqs. (5.16) through (5.18) involves only the odd values of N, the number of complex coefficients, A_j, is $N + 1$ and the number of real coefficients, a_j and b_j, is $2(N + 1)$.

These coefficients can be determined from thermoelastic data, S^*. It should be noted that by using conformal mapping and analytic continuation techniques, Eqs. (5.16) through (5.18) imply that the induced stresses satisfy equilibrium and traction-free conditions in the adjacent portion of the entire boundary. However, unlike a classical boundary-value problem where one would typically evaluate the unknown coefficients, A_j, by satisfying the boundary and loading conditions around the entire shape, one can use a combination of the measured stresses and/or displacements from within region R_z to determine these unknown complex coefficients, A_j. Additional known boundary conditions may also be imposed at discrete locations. The concept of collecting measured data in a region R^* adjacent to an edge Γ, mapping R_z into R_ζ such that Γ of the physical z-plane is mapped into the unit circle in the ζ-plane whereby the traction-free conditions on Γ are satisfied continuously, relating the two complex stress functions to each other, plus satisfying other loading conditions discretely on the boundary of the component beyond Γ will be referred to as the *mapping-collocation technique*.

The interior load induced thermal information S^* at m different locations within region R^* are employed. A system of simultaneous linear equations $[S]_{m\times 2(N+1)}\{c\}_{2(N+1)\times 1} = \{S^*\}_{m\times 1}$, is formed whose matrix $[S]$ consists of analytical expressions of TSA signals S^*, Eq. (5.19), vector $\{c\} = \{a_{-N}, b_{-N}, a_{-N+2}, b_{-N+2}, \ldots, a_{N-2}, b_{N-2}, a_N, b_N\}$ has $2(N+1)$ unknown real coefficients, and vector $\{S^*\}$ includes the m measured TSA signal values of S^* such that $m \gg 2(N+1)$. The best values of the coefficients A_j, in a least-squares numerical sense, are then be determined. The variables $\zeta_j = \xi + \mu_j\eta$ are related to the physical locations $z = x + iy$ through the inverse mapping function $z_j = \omega_j(\zeta_j)$ of Eqs. (5.12) through (5.13).

5.4 Inverse Method Procedure

The particular inverse method used here is combining TSA signal produced from Airy stress function scheme with load induced thermal information measured by means of TSA. Through an iterative process that determines new constitutive parameters, the difference between measured TSA signals and the ones produced from Airy stress function is minimized. The function to be minimized is

$$f\left(\widehat{S}_{Airy}, P\right) = \|r\|, \quad \text{where } r = \widehat{S}_{TSA} - \widehat{S}_{Airy} \tag{5.20}$$

where $\widehat{S}_{Airy}$ and $\widehat{S}_{TSA}$ are vector containing thermoelastic signals data determined by Airy stress function scheme and TSA respectively. P is a vector containing the constitutive parameters, $E_1, E_2, \nu_{12}, G_{12}, K_1, K_2$ and $\|r\|$ is the norm of r. Because Eq. (5.20) is nonlinear with respect to P, iterative procedures are appropriate methods for minimizing of $f\left(\widehat{v}_{Airy}, P\right)$ and determination of P. LMA (Levenberg-Marquardt Algorithm) is commonly used because it combines the benefits of Steepest Descent Method and Gauss-Newton Method. The LMA has the form [14]

$$P_{i+1} = P_i - \left(J^T J + \lambda\cdot \operatorname{diag}\left(J^T J\right)\right)^{-1} J^T r \tag{5.21}$$

where i is iteration number, J and J^T are Jacobian and Jacobian transpose, determined by backward difference, $J_{m,n} = \frac{\partial r_m}{\partial P_n}$; m is number of nodal displacements and n is number of constitutive parameters (6 in this work), and λ is non-negative damping factor, adjusted each iteration step, adjusts between Steepest Descent Method and Gauss-Newton Method.

The primary disadvantage of LMA is the need for matrix inversion during each iteration. In most applications, reduced iterations compensate for the matrix inversion. After calculating a new P_{i+1}. The constitutive parameters are checked for validity, i.e., a positive-definite stiffness matrix, and are adjusted if not valid. The validated P_{i+1} are inputs to a new analysis and the resulting nodal displacements are used to determine f_{i+1}. If $f_{i+1} < f_i$, the constitutive parameters are updated, $P_{i+1} \to P_i$, λ is reduced by a factor of 10, and the next iterations begins. If $f_{i+1} > f_i$, then λ is incrased by a factor of 10 and P_i is not updated. As $\lambda \to 0$, LMA becomes exactly the Gauss-Newton Method.

5.5 Numerical Experiment

The developed inverse hybrid-TSA approach is utilized to analyze a finite-width tensile $[0_{13}/90_5/0_{13}]$ graphite/epoxy orthotropic plate ($E_1 = 101$ GPa, $E_2 = 24.9$ GPa, $G_{12} = 2.88$ GPA, $\nu_{12} = 0.152$ similar to the one used in Ref. [15]) with circular hole of radius $R = 6.35$ mm. The plate was loaded in the stiffest material direction (1-, x-direction), Fig. 5.3. Over-all laminate dimensions are 381 mm long and 38.1 mm wide. The coordinate origin is at the center of the plate and the response is symmetric about x- and y-axes. The thermoelastic coefficients K_1 and K_2 were evaluated from uniaxial tensile

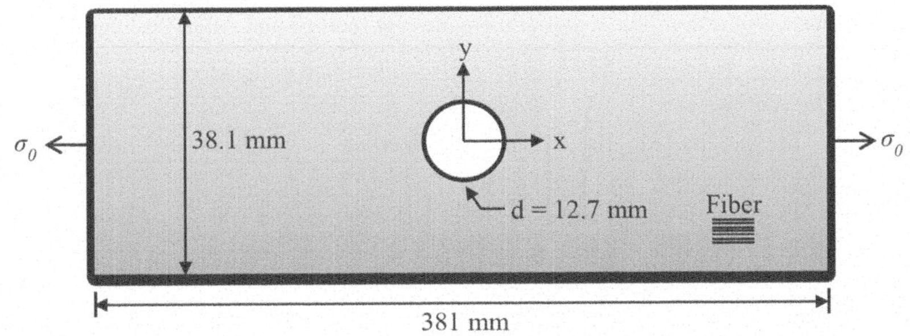

Fig. 5.3 Schematic of symmetrically-loaded finite Gr/E $[0_{13}/90_5/0_{13}]$ composite plate with central circular hole

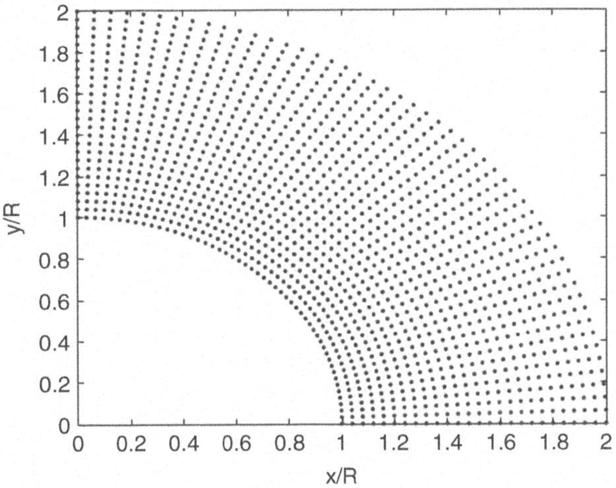

Fig. 5.4 Locations of m = 1326 data points used in the numerical experiment

coupons loaded in the stiff and compliant laminate orientations of the orthotropic $[0_{13}/90_5/0_{13}]$ composite plate. The values of the thermomechanical coefficients were determined to be $K_1 = 1.8$ mU/MPa (12.38 U/psi) and $K_2 = 14.7$ mU/MPa (101.25 U/psi). The unit U is used to signify the raw TSA output, in uncalibrated signal units [15].

Using simulated data as input is not a representation of the real situation when employing measured data and therefore not a sufficiently severe test of the present scheme. Pseudo-experimental errors were therefore imposed on the simulated input according to the following equation in order to simulate the scatter and/or error typically associated with measured information

$$S_i^* = (1 + E R_S)\, S_{i,s}^*$$

where S_i^* are simulated measured-input TSA signals at location *i*, E is the maximum absolute random error (user specified), R_S are independent random numbers $(-1 \le R_S \le 1)$ and $S_{i,s}^*$ are 'error-free' values of the simulated TSA signals at the position *i*. The random errors of these TSA signals were independently generated at each input location. This was accomplished by evaluating R_S using a random number generation program. The maximum absolute error E at a location *i* is a controlled percentage of the simulated TSA signals $S_{i,s}^*$ at that location. For this numerical experiment, an absolute error of $E = 20\%$ was used. Using these 'noisy' input values at the 1326 locations of Fig. 5.4, the constitutive properties were determined and compared with the values used for the simulated measured-input displacement.

5.6 Results

5.6.1 Evaluating Constitutive Properties

The results of this numerical experiment, based on 1326 input values of S^* distributed as shown in Fig. 5.4, were used to evaluate 2 complex coefficients, A_k, (4 real coefficients). Table 5.1 shows the predicted constitutive properties using different

Table 5.1 Identified constitutive parameters of orthotropic material using inverse method and Airy stress function scheme

	Initial guesses						Identified parameters					
	E_1 (GPa)	E_2 (GPa)	G_{12} (GPa)	ν_{12}	K_1	K_1	E_1 (GPa)	E_2 (GPa)	G_{12} (GPa)	ν_{12}	K_1	K_2
Case (a)	171.7	39.84	1.728	0.182	18.57	91.125	105.8	25.91	2.802	0.154	12.71	100.7
	70%	60%	40%	20%	50%	10%	4.75%	4.07%	2.71%	1.36%	2.74%	0.55%
Case (b)	90.9	17.43	1.728	0.167	18.57	111.38	98.84	23.30	2.63	0.155	13.71	103.4
	10%	30%	40%	10%	50%	10%	2.14%	6.43%	8.58%	2.14%	10.7%	2.14%
Case (c)	80.8	27.39	3.74	0.122	13.62	21.5	105.8	24.3	2.68	0.159	12.1	96.5
	20%	10%	30%	20%	10%	20%	4.72%	2.36%	7.08%	4.72%	2.36%	4.72%
Case (d)	181.8	2.49	2.02	0.274	16.09	121.5	108.9	22.72	2.80	0.164	12.74	1.95
	80	90	30	80	30	20	7.78%	8.76%	2.92%	7.78%	2.92%	1.96%

Target values of elastic constants are $\boldsymbol{E_1} = 101$ GPa, $\boldsymbol{E_2} = 24.9$ GPa, $\boldsymbol{G_{12}} = 2.88$ GPa, $\boldsymbol{\nu_{12}} = 0.152$, $\boldsymbol{K_1} = 12.38$ U/psi, $\boldsymbol{K_2} = 101.25$ U/psi [15]

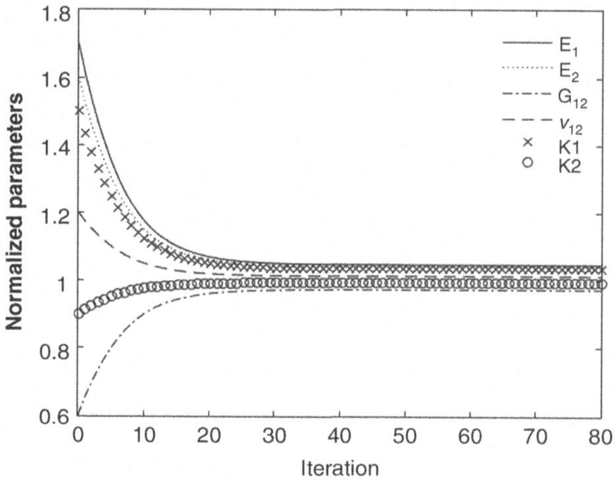

Fig. 5.5 The convergence of the proposed inverse method for case (a)

initial guesses; E_1^0, E_2^0, G_{12}^0, ν_{12}^0, K_1^0, K_2^0. Regardless of the initial guesses and the noisy implemented data, the method converges to the true value within 20 iterations. All the constitutive properties similarly converged as shown in Fig. 5.5.

5.6.2 Evaluating Stresses

Contour plots of the normalized longitudinal, transverse, and shear stresses in the region adjacent to the circular hole are shown in Fig. 5.6. The results shown in Fig. 5.6 demonstrate an excellent agreement with the simulated data. The results of this numerical experiment based on the discrete input values of $S*$ agree virtually exactly with the FE-simulated values throughout region R_z. The maximum error of the predicted maximum values of σ_{xx} agrees within 1% of the maximum FE-predicted value. The results for σ_{yy} and σ_{xy} are similarly excellent. The u-displacement in Fig. 5.6a is also excellent even though it was derived from Airy stress scheme. These results illustrate the ability of the present hybrid inverse method to provide reliable stresses even with such bad or noisy input data and without the knowledge of the constitutive properties and TSA calibration factors. Such numerical experiments employing simulated test data from FE solution help verifying that there are no algebraic errors and substantiate that the system is numerically robust.

5.7 Summary, Discussion and Conclusions

Stress analysis of materials can be accomplished by the determination of stress concentrations within structures. Determination such factors necessitates the need of using experimental techniques and knowing the constitutive properties of the structure material and TSA calibration factors. Stress analysis of any structure using only TSA signal data cannot be accomplished without prior knowledge of the constitutive properties of the within material and TSA calibration coefficients,

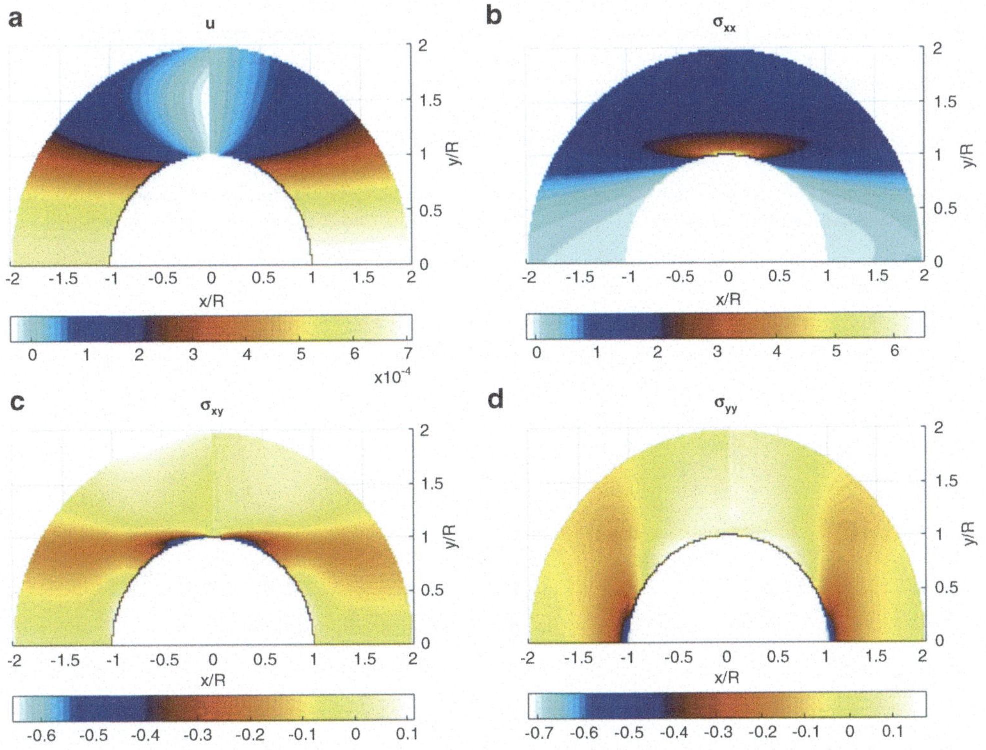

Fig. 5.6 Contour plot of (**a**) u-displacement, (**b**) σ_{xx}/σ_0, (**c**) σ_{yy}/σ_0, and (**d**) σ_{xy}/σ_0 from hybrid scheme (left) and ANSYS (right)

so long as an Airy Stress Function exists to describe the geometry and load configuration. A new inverse hybrid method which processes the load-induced TSA signals with a stress function in complex variables, together with conformal mapping and analytic continuation concepts, provides the constitutive parameters, including TSA calibration factors, and the individual stresses on and in the neighborhood of a circular hole in a finite orthotropic composite plate, all from a single test. The new inverse problem formulation is developed using the Airy stress function, Levenberg-Marquardt Algorithm, and TSA signal data to determine the constitutive properties of a graphite/epoxy composite loaded in the strong/stiff direction. The primary advantage of this new formulation is the direct use of TSA signals data to determine constitutive properties as well as separating the stresses into three individual components. The inverse method algorithm determined the constitutive properties with errors less than 10%.

References

1. Lin, S.T., Rowlands, R.E.: Thermoelastic stress analysis of orthotropic composites. Exp. Mech. **35**(3), 257–265 (1995)
2. Alshaya, A., Shuai, X., Rowlands, R.: Thermoelastic stress analysis of a finite orthotropic composite containing an elliptical hole. Exp. Mech.**56**(8), 1373–1384 (2016)
3. Hawong, J.S., Lin, C.H., Lin, S.T., Rhee, J., Rowlands, R.E.: A hybrid method to determine individual stresses in orthotropic composites using only measured isochromatic data. J. Compos. Mater. **29**(18), 2366–2387 (1995)
4. Alshaya, A., Rowlands, R.: Experimental stress analysis of a notched finite composite tensile plate. Compos. Sci. Technol. **144**, 89–99 (2017)
5. Baek, T.H., Rowlands, R.E.: Experimental determination of stress concentrations in orthotropic composites. J. Strain Anal. Eng. Des. **34**(2), 69–81 (1999)
6. Baek, T., Rowlands, R.: Hybrid stress analysis of perforated composites using strain gages. Exp. Mech. **41**(2), 195–203 (2001)
7. Avril, S., Pierron, F.: General framework for the identification of constitutive parameters from full-field measurements in linear elasticity. Int. J. Solids Struct. **44**(14–15), 4978–5002 (2007)
8. Alshaya, A., Considine, J.M., Rowlands, R.: Determination of constitutive properties in inverse problem using airy stress function. In: Baldi, A., Considine, J.M., Quinn, S., Balandraud, X. (eds.) Residual Stress, Thermomechanics & Infrared Imaging, Hybrid Techniques and Inverse Problems, Volume 8: Proceedings of the 2017 Annual Conference on Experimental and Applied Mechanics, pp. 73–81. Springer International Publishing, Cham (2018)
9. Lekhnitskii, S.G.: Anisotropic Plates. Gordon & Breach Scientific Publishers, New York (1968)
10. Muskhelishvili, N.: Some Basic Problems of the Mathematical Theory of Elasticity, 1977 edition. Springer, Leyden (1977)

11. Savin, G.N.: Stress Concentration Around Holes. Pergamon Press, New York (1961)
12. Gerhardt, T.D.: A hybrid/finite element approach for stress analysis of notched anisotropic materials. J. Appl. Mech. **51**(4), 804–810 (1984)
13. Huang, Y.-M.: Determination of individual stresses from thermoelastically measured trace of stress tensor. PhD Thesis, University of Wisconsin-Madison, Madison (1989)
14. Considine, J.M., Vahey, D.W., Matthys, D., Rowlands, R.E., Turner, K.T.: An inverse method for analyzing defects in heterogeneous materials. In: Proulx, T. (ed.) Application of Imaging Techniques to Mechanics of Materials and Structures, Volume 4: Proceedings of the 2010 Annual Conference on Experimental and Applied Mechanics, pp. 339–346. Springer New York, New York (2013)
15. Alshaya, A. A.: Experimental, analytical and numerical analyses of orthotropic materials and biomechanics application. PhD Thesis, University of Wisconsin-Madison, Madison (2017)

Chapter 6
Experimental Investigation of Emissivity Influence to Obtain Thermal Field by Near Infrared Thermography

C. Zhang, J. Marty, A. Maynadier, P. Chaudet, J. Réthoré, and M.-C. Baietto

Abstract Thermal fields are usually obtained by infrared camera (operating in 3–12 μm). However, the infrared cameras are expensive, fragile and low resolution, thus the infrared camera is more used for laboratory researches. Nowadays, silicon-based sensor camera has been widely used to perform real-time observation of the kinematic fields, mainly thanks to digital image correlation or interferometry. Silicon-based camera is also sensitive in the near infrared spectral ranges (operating in 0.7–1.1 μm). An automatic prediction of the camera exposure time is developed and presented in this paper to follow the surface emissivity of the sample and avoid saturation of the gray level.

Keywords Thermography · Emissivity · Silicon camera · Radiometric model · Near infrared

Kinematic and thermal field measurements can provide rich information in the field of thermo-mechanics of materials and structures (e.g., constitutive model identification, fatigue behavior investigation, cracking detection, etc.). Up to now, silicon-based sensor cameras (CCD and CMOS) have been widely used to perform in situ observation of the kinematic field on the material surfaces, mainly thanks to digital image correlation or interferometry. Measuring temperature distribution of an object undergoing thermal cycle is also very important and meaningful in the field of thermo-mechanics of materials. To date, thermal fields are usually obtained by mid-wave infrared camera (operating in 3–12 μm), which can transform the thermal energy, emitted by objects in the infrared band of the electromagnetic spectrum (i.e. from 1 to 1000 μm), into an electronic video signal. However, the infrared camera is more used for laboratory researches because it is expensive, fragile, low resolution and difficult to use. The silicon-based camera is also sensitive in the near infrared spectral ranges (operating in 0.7–1.1 μm), thus it is possible and suitable for silicon-based camera to acquire the thermal fields at high temperature. Compared with infrared camera, silicon-based camera is low-cost, low noise, durable and high resolution, and can be widely used in industrial applications. In the near infrared spectral ranges, a small temperature variation readily cause a large modification in the image illumination (gray level), readily resulting in the acquired images with saturation or poor dynamic range of gray levels. The bad image acquired will make the thermal field measurement impossible.

In previous work we have proposed a method to precisely and automatically adjust the exposure time of the camera to control the image illumination whatever the temperature evolution is. The principle is to use the two last image to predict the next exposure time of the camera to keep the gray level constant when the camera frequency is maintained constant. In our study, the radiometric model of camera describing the relationship between intensity and temperature is given by [1]:

$$I_n(T) = \frac{gl}{\tau}(T) = k_w \exp\left(\frac{-C_2}{\lambda_x(T)T}\right) \tag{6.1}$$

C. Zhang (✉) · P. Chaudet · M.-C. Baietto
LaMCos, Université de Lyon/INSA Lyon/UMR CNRS 5259, Villeurbanne Cedex, France
e-mail: chao.zhang@insa-lyon.fr

J. Marty (✉)
LaMCos, Université de Lyon/INSA Lyon/UMR CNRS 5259, Villeurbanne Cedex, France

ESTA LAB, Belfort, France
e-mail: jeremy.marty@insa-lyon.fr

A. Maynadier
University of Bourgogne Franche-Comté, FEMTO-ST Institute, CNRS/UFC/ENSMM/UTBM, Department of Applied Mechanics, Chemin de L'épitaphe, Besançon, France

J. Réthoré
GeM, Ecole Centrale de Nantes/Université de Nantes/UMR CNRS 6183, Nantes, France

A. Baldi et al. (eds.), *Residual Stress, Thermomechanics & Infrared Imaging, Hybrid Techniques and Inverse Problems, Volume 7*, Conference Proceedings of the Society for Experimental Mechanics Series, https://doi.org/10.1007/978-3-319-95074-7_6

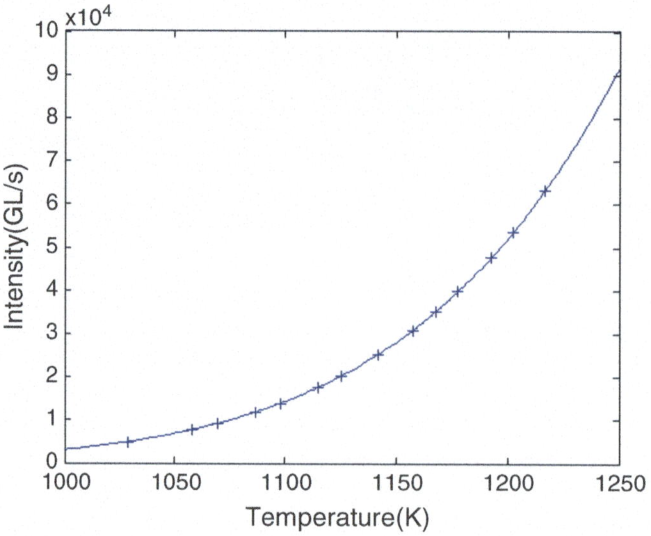

Fig. 6.1 Experimental data and radiometric calibration curve

where gl, τ and I_n are respectively the gray level, the exposure time and the intensity of the image and T the temperature. C_2 is the Planck constant equal to 1.44×10^{-2} m · K, $\lambda_x(T)$ is the extended effective wavelength and is defined in the following Equation [2]:

$$\frac{1}{\lambda_x(T)} = a_0 + \frac{a_1}{T} + \frac{a_2}{T^2} + \frac{a_3}{T^3} + \cdots + \frac{a^n}{T^n}, (n = \infty) \tag{6.2}$$

where $a_0, a_1, a_2, a_3, \ldots, a_n$ and the parameter k_w of Eq. (6.1) are parameters of the camera which should be determined by radiometric calibration process. The radiometric model of our camera is plotted on the Fig. 6.1 and the parameters are given in the tabular.

The prediction of the exposure time is based on the Planck law (1) where the extended wavelength has only one parameter a_0. During the acquisition of two images, we suppose that the camera frequency is enough high to have a small difference of temperature (magnitude of 10° or 20°). So the extended wavelength is depending of only one parameter. Thus the prediction of the next intensity of image is equal to [3]:

$$I_n(i) = \frac{\overline{gl(i)}}{\tau(i)} = \frac{I_n(i-1)^2}{I_n(i-2)} = \left[\frac{\overline{gl(i-1)}}{\tau(i-1)}\right]^2 \times \frac{\tau(i-2)}{\overline{gl\left(i-2\right)}} \text{ with i} > 2 \tag{6.3}$$

In order to make sure the gray level constant (i.e., that $\overline{gl(i)} = \overline{gl(1)}$), the next exposure time can be predicted as follows:

$$\tau(i) = \overline{gl(1)} \times \left[\frac{\tau(i-1)}{\overline{gl(i-1)}}\right]^2 \times \frac{\overline{gl(i-2)}}{\tau(i-2)} \tag{6.4}$$

These experiments have been realized with a black body device which has a surface emissivity uniform and equal to one. The thermal field is easily obtained by inverting the radiometric model of the optic device.

However a specimen surface has heterogeneities of emissivity due to the material or even the coating if you add one on the surface. The Planck law presented before in Eq. (6.1) is only valid with an emissivity equal to one which is the case of a black body. In practical application, the specimen surface has heterogeneities of emissivity and the measurement of different surface emissivity is more difficult. In this study, we propose a novel method to measure the thermal fields of specimen surface with different emissivity by using silicon-based camera. Firstly, a specimen surface is sprayed with paint with different emissivities, which were made by mixing different proportions of white paint and black paint. Secondly, several thermocouples are welded to specimen surface to measure the temperature of various regions with different emissivities. Thirdly, the Planck's algorithm proposed above is used to predict new exposure time to control the mean image gray level constant with temperature evolution from 873 to 1023 K. Fourthly, all radiometric models of various regions with different emissivities can be calibrated by experimental data. This method will give the evolution of emissivity on the radiometric model and will permit to perfectly understand the evolution of temperature in function of the emissivity of a specimen.

References

1. Orteu, J.J., Rotrou, Y., Sentenac, T., Robert, L.: An innovative method for 3-D shape, strain and temperature full-field measurement using a single type of camera: principle and preliminary results. Exp. Mech. **48**, 163–179 (2008)
2. Saunders, P.: General interpolation equations for the calibration of radiation thermometers. Metrologia. **34**, 201–210 (1997)
3. Zhang, C., Marty, J., Maynadier, A., Chaudet, P., Réthoré, J., Baietto, M.C.: A novel approach to control gray level of image invariant during near infrared thermography. 23ème Congrès Français de Mécanique-S04 Mesures de champs et mécanique expérimentale (2017)

Chapter 7
Experimental Validation of the Energy Balance Equation in the Presence of Acoustic Emission

Natasha Bradley, Mulugeta Haile, Benjamn Northington, Michael Coatney, and Amber Eure

Abstract This study attempts to experimentally validate the energy balance equation during a quasi-static loading of polymer composites with explicit consideration of the energy dissipated through acoustic emission (AE). The experimental protocol estimates the various terms of the energy balance equation, namely the external work done E_W, the elastic stored energy E_ε, the dissipated heat energy E_H, and the dissipated acoustic emission energy E_{AE} using in-situ measurements. The elastic stored energy E_ε is derived from strain gage data, the dissipated heat energy E_H is estimated from the temperature field obtained from quantitative infrared thermography (IT) and surface mounted thermocouples, and the acoustic emission energy E_{AE} is approximated using two surface mounted piezoelectric transducers. The total input energy E_w, i.e. the external work done by the loading system is calculated from the cross-head displacement. The energy balance equation, ignoring inertial forces, is written as $E_W - E_\varepsilon - E_H - E_{AE} = Constant$. The energy approach presented herein may be used to quantify the level of material degradation as well as remaining useful life in primary load carrying structures.

Keywords Acoustic Emission · Energy Balance · AE Energy · Crack Energy · Thermography

7.1 Materials and Methods

The experimental setup, shown in Fig. 7.1a, includes (1) mechanical testing system (MTS), (2) and flash thermography system (EchoTherm™), (3) an acoustic emission sensing system (Vallen™), (4) strain gages, (5) thermocouples and, (6) continuous data acquisition module (LabView). The test specimen is a commercially available ±45 textile carbon-fiber polymer composite (McMaster-Carr). The thermography unit includes a highly sensitive and high speed infrared camera (FLIR) and two Xeon bulbs that emit burst of flash light. It is controlled by a software (Virtuoso™) that calculates thermal diffusivity and related thermophysical properties using the rate of change of surface temperature after the application of the flash light [1]. The acoustic emission system includes two 9.5 mm by 11.8 mm broadband piezoelectric transducers with a uniform frequency response in the range 60–850 kHz and pre-amplifiers with software controllable gain of 34 dB. Strain gage and thermocouple data are recorded using LabView compact DAQ acquisition module. The LabView system is also used to synchronize data recording from various devices and sensors [2]. The rectangular test specimen, shown in Fig. 7.1b, has length, $l = 150$ mm, width $w = 25$ mm, and thickness $t = 3.2$ mm. The gage area is $l = 80$ mm and $w = 15$ mm with same thickness. Figure 7.1a also shows the loading of the test specimen and a dummy sample (unloaded specimen), the latter is used as a reference during thermography analysis.

An incremental ramp-and-hold quasi-static test was conducted while continuously recording data from the various sensors. The hold, typically less than 5 min, was required to synchronize the thermography data with other measurements as well as to examine the state of the test specimen at various strain levels. The ramp-and-hold sequence is programmed with constant 2.5 kN load increments, which is 5% of the breaking load of the test specimen. The test was conducted until complete fracture at 50 kN.

N. Bradley (✉) · M. Haile · B. Northington · M. Coatney · A. Eure
U.S. Army Research Laboratory, Vehicle Technology Directorate, Aberdeen Proving Ground, MD, USA
e-mail: natasha.c.epps-bradley.civ@mail.mil

A. Baldi et al. (eds.), *Residual Stress, Thermomechanics & Infrared Imaging, Hybrid Techniques and Inverse Problems, Volume 7*, Conference Proceedings of the Society for Experimental Mechanics Series, https://doi.org/10.1007/978-3-319-95074-7_7

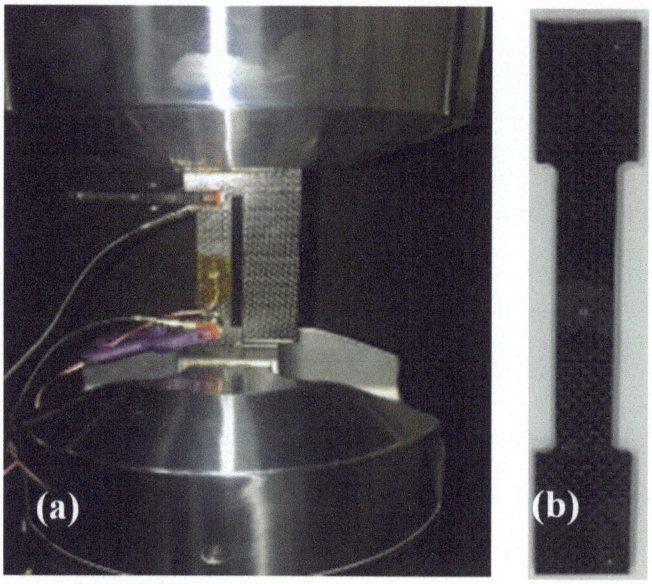

Fig. 7.1 (**a**) Experimental setup showing loading frame and sensors. (**b**) Geometry of the test specimen

7.2 Experimental Results

Figure 7.2 (left axis) shows the stress-strain behavior of the test specimen; a superimposed AE energy vs strain is also plotted on the right axis. AE activity is attributed to the rapid release of energy, typically due to microscopic material separation or plastic deformation [3]. The energy propagates through the continuum solid as Rayleigh wave and recorded using surface mounted piezoelectric transducers. The electrical signal strength of the piezo is directly proportional to the strength and intensity of the event that causes the emission [4]. Hence, the energy of AE event can be estimated from the energy of the transducer waveform. Similar to an audio wave or other 1D electrical signals, the energy of the transducer waveform may be estimated from the square of the voltage. So in a simplified analysis one can estimate the energy by integrating the square of the transducer output over the life of the signal. In the test presented here, two transducers are placed near the fillet symmetrically about the mid-section of the specimen. The transducers are sensitive to record acoustic emissions up to 1 MHz, which is broad enough to acquire most of the AE energy over a wide spectrum [5]. The energy of the transducer signal is calculated by:

$$E_{AE} = \int_{to}^{t} v_i(t)^2 dt \tag{7.1}$$

Here $v_i(t)$ is the transient transducer output, t_o is starting time at threshold crossing, and t is the end time of the signal. The square summable sequence shown in Eq. (7.1) only provides an estimate of the event energy or intensity. A more accurate formulation would require the material properties of the test specimen (such as E-modules, density, attenuation, and dispersion), volume of propagation, boundary conditions, as well as the size of the piezo element used to transduce the surface vibration into electrical voltage. Nonetheless, the estimations are still directly correlated to the actual event energy, despite loosely, and likely with a low order polynomial. In order to use the estimates of Eq. (7.1) within the framework of the conservation of energy, such as the energy balance equation, it has to be converted into appropriate energy unit. However, conversion of AE signal energy into an appropriate SI energy unit, such as Joule (J), requires extensive calibration of transducer response at various conditions and is not part of this report. Rather, an energy unit (Eu) is adapted for the AE energy simultaneously limiting the focus on the evolution of individual term in the balance equation $E_W - E_\varepsilon - E_H - E_{AE} = Constant.$

A portion of the work done by the loading frame is stored as an elastic (or strain) energy in the material. The elastic stored energy, E_ε, of a rectangular bar stretched from its initial length l_1 to length l_2 is given by:

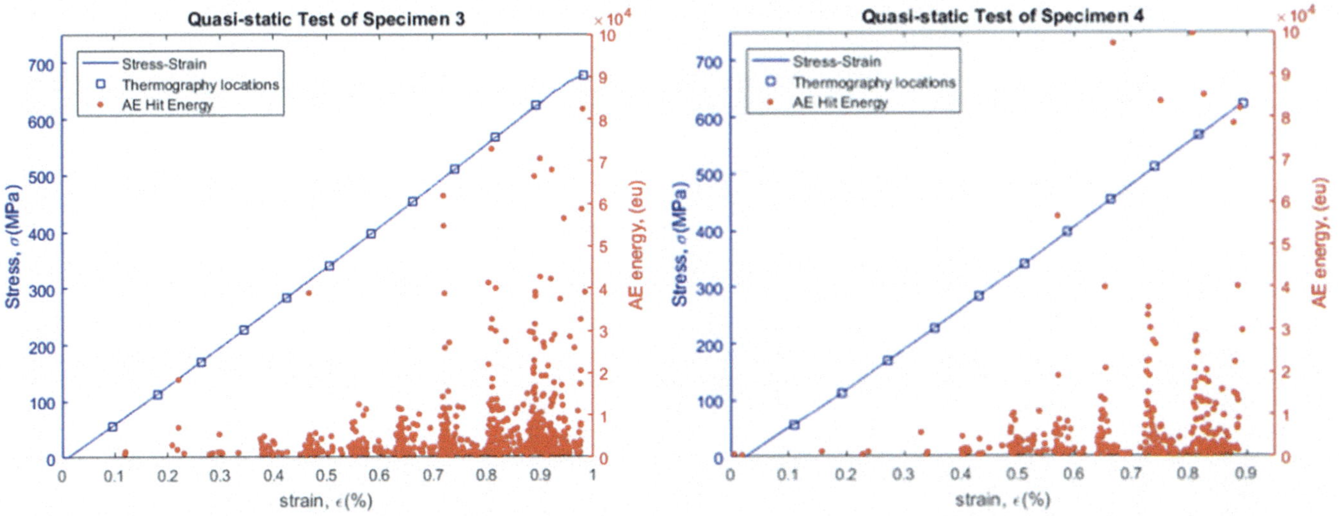

Fig. 7.2 Stress and AE energy are plotted against applied strain in a displacement control quasi-static test setting for two nominally identical test specimens. Blue rectangular dots indicate the instances of flash thermography. The red dots indicate the calculated AE energy of each event in energy units (eu). The distinct spike in AE count corresponds to ramp cycles

$$E_\varepsilon = \int_{l_1}^{l_2} \frac{P(l)}{2EA} dl \tag{7.2}$$

Where E is the elastic modulus, P(l) is the applied (MTS) load and A is the cross-sectional area of the test specimen. The stored energy can also be approximated from the area under the stress-strain curve, where the cumulative stored energy at a given strain level ε_i will be $0.5(\sigma_i\varepsilon_i)$ and the incremental value at ramp cycle ε_{i+1} will be the difference [6]. For the specimens tested here, the stress –strain response remains linear all the way to failure.

The external work done by the loading system, E_w, per ramp cycle, is estimated by:

$$E_w = \int_{l_1}^{l_2} P(l) dl \tag{7.3}$$

Where P(l) is axial load, l_1 and l_2 are gage lengths at the beginning and end of the ramp cycle. The heat dissipation (or the unrecoverable thermal energy) is estimated by:

$$E_H = mc\,(T_2 - T_1) \tag{7.4}$$

Where m is mass (of the gage area), c is the specific heat, T_1 and T_2 are the specimen temperatures at the beginning and end of each ramp cycle measured by surface mounted thermocouples. The energy balance equation for each ramp cycle can now be written as:

$$E_W - E_\varepsilon - E_H - E_{AE} = Constant \tag{7.5}$$

7.3 Summary

Mechanical loading results in stored elastic energy, release of AE, and heat. In general, AE activity is directly related to the materials ability to store elastic energy E_ε. In the early phase of loading, the material typically stores all the external work done as an elastic energy. There is almost no AE activity and not much change in surface temperature. However, as the applied load (or displacement) increases, the specimen's ability to store additional energy reaches its limit and AE, typically small energy, starts to appear. As the applied load continues to increase further, the material releases most if the input energy in the form of high energy AE waves and heat. The energy of the AE waves continue to grow until the material completely loses its ability to store strain energy. The journal version of this report contains detail assessments of the energy terms in Eq. (7.5) including plots of each terms as a function of applied load for various test samples. The energy approach presented herein may be used to quantify the level of material degradation as well as remaining useful life of primary load carrying structures.

References

1. Haile, M.A., Bradly, N.C., Coatney, M.D., Hall, A.J.: Detection of early stage material damage using thermophysical properties. In: Baldi, A., Considine, J., Quinn, S., Balandraud, X. (eds.) Residual Stress, Thermomechanics & Infrared Imaging, Hybrid Techniques and Inverse Problems, Volume 8. Conference Proceedings of the Society for Experimental Mechanics Series. Springer, Cham (2018)
2. Haile, M.A., Chen, T.K., Sediles, F., et al.: Estimating crack growth in rotorcraft structures subjected to mission load spectrum. Int. J. Fatigue. **43**, 142–149 (2012)
3. Haile, M.A., Hall, A.J., Yoo, J.H., et al.: Detection of damage precursors with embedded magnetostrictive particles. J. Intell. Mater. Syst. Struct. **27**, 1567–1576 (2015)
4. Haile, M.A., Riddick, J.C., Assefa, A.H.: Robust particle filters for fatigue crack growth estimation in rotorcraft structures. IEEE Trans. Reliab. **65**, 1438–1448 (2016)
5. Haile, M., Ghoshal, A.: Application of compressed sensing in full-field structural health monitoring. In: Proceedings of the SPIE 8346, Smart Sensor Phenomena, Technology, Networks, and Systems Integration, San Diego, pp. 46–53 (2012)
6. Haile, M.A., Chen, T.-K., Shiao, M., Le, D.: Crack growth behavior in preloaded metallic nested-angle plates under flight load spectrum. Exp. Appl. Mech. **6**, 3–11 (2011)

Chapter 8
Model Based Inversion for Pulse Thermography

Stephen D. Holland

Abstract We present a model-based inversion approach to the interpretation of flash thermography data. Traditionally, flash thermography has been image sequences have been hard to interpret in the presence of lateral heat flows. This is a particularly significant problem for composite laminate materials where heat conducts readily in the fiber planes. We use a linear inversion approach to represent the observed surface temperature as a superposition of contributions from buried reflectors. We investigate the additional complexities of considering cases where the specimen is curved, changing the nature of the heat flow. By solving a large linear inversion problem we are able to determine the strengths of the buried reflectors and create a concrete representation of observed defects. Results from this approach are compared to more traditional analysis approaches for flash thermography.

Keywords Flash thermography · Nondestructive evaluation (NDE) · Inversion algorithms · Model-based inversion

8.1 Introduction

Pulse thermography uses heat flows from a flash or pulse to find flaws in materials by observing the cool down of the exposed surface with a thermal camera [1]. A thick material takes longer to equilibriate than a thin material, so the cool-down profile will be different. The most widely used analysis technique is thermographic signal reconstruction (TSR) [2]. TSR is a method for fitting the observed data to curves, which can then be differentiated. The second derivative of the log temperature vs. log time graph indicates the changing dimensionality of the heat flow, from one-dimensional (into the specimen thickness) to higher dimensional flows around a defect and/or simple equilibriation. TSR gives a clear quantitative physical interpretation of thickness when the heat flows are purely one-dimensional. Once geometry is more complicated or if small defects are present the interpretation is qualitative.

This presentation illustrates an quantitative interpretation of pulse thermography data based on a Green's function heat flow model [3]. We represent measured surface temperature as a linear combination of the excitation pulse plus "reflections" from buried insulators. A regularizes linear inversion is used to solve for effective source intensities for the insulators. Source intensity maps at multiple depths give a concrete interpretation of defect locations and sample geometry based on the pulse thermography data.

8.2 Methods

The buried insulators are modeled as buried heat sources that act to partially or completely counterbalance the downward heat flow from the excitation pulse. The response at the surface to such a buried insulator is the convolution of the spatial (z) derivative of the 1D Green's function for heat flow of the pulse from the surface to the insulator with the 3D Green's function for heat flow from the insulator back to a surface measurement point. Let b represent the measured surface temperature, unwrapped over space and time into a column "vector". Let x be a vector with elements representing the intensities of the pulse source and all the buried reflectors. The measured surface temperatures can then be modeled as a system of equations

$$Ax = b$$

S. D. Holland (✉)
Iowa State University, Ames, IA, USA
e-mail: sdh4@sdholland.com

A. Baldi et al. (eds.), *Residual Stress, Thermomechanics & Infrared Imaging, Hybrid Techniques and Inverse Problems, Volume 7*, Conference Proceedings of the Society for Experimental Mechanics Series, https://doi.org/10.1007/978-3-319-95074-7_8

where each column the matrix A represents the surface temperature-time response to the corresponding source intensity entry in x. The columns in A can be approximated from the convolutions of known Green's function solutions. By inverting the system it is possible to estimate source intensities x from the measured data b.

There are a host of practical difficulties in performing such an inversion. For a measurement surface of any size with a reasonable number of frames, b can be huge. In our prototype implementation we break the surface into 24 mm by 22 mm tiles and solve each tile independently. The "vector" b is resulting 48 by 44 by 367 frame image sequence with a total of approximately 800,000 numbers. There are also 1000–2000 entries in x, so A has on the order of a billion elements. Both constructing and inverting A are non-trivial computational problems.

Nevertheless with modern computers, such a large inversion is feasible. The various other challenges, such as lateral boundaries and minimizing noise through regularization are also solvable. The result of the inversion process are the elements of x, representing pulse source intensity amplitudes and the reflector amplitudes for the buried layers.

8.3 Results

The method was applied to experimental data from carbon fiber reinforced polymer (CFRP) specimens. Two examples are shown, the first from a specimen with flat-bottom holes and the second from a specimen with impact damage. Figure 8.1 shows model-based inversion results for a specimen with flat-bottom holes alongside a corresponding TSR second derivative image. The model-based inversion is able to detect the holes at this depth (the second two rows) except for the smallest size. Thermographic signal reconstruction is slightly more sensitive, but the indications are significantly blurred laterally. The inversion process (as would be expected) causes noise gain, but compresses the hole indications to their true sizes. The model-based inversion colormap is calibrated in kJ/m^2 of buried source intensity (compared to the incident pulse of 11.7 kJ/m^2).

Figure 8.2 shows a slice at 1.17 mm depth of a CFRP specimen subject to impact damage. The characteristic "S"-shape of impact delamination damage is shown. Some regions, hidden behind shallower delaminations, show as dark (negative) reflection intensities, which are obviously non-physical. Delaminations at this depth include the tips of the "S" and a faint fan-shape on the lower-left side of the impact.

8.4 Conclusions

Model based inversion for pulse thermography provides a concrete interpretation of pulse thermography measurements, even in the presence of lateral diffusion. The measured data is represented as the initial pulse plus a sum of responses from buried reflectors (heat sources) at several depths. The reflector intensities can be determined through regularized linear inversion. The method has been illustrated on two carbon fiber specimens, one with flat bottom holes and one with impact damage.

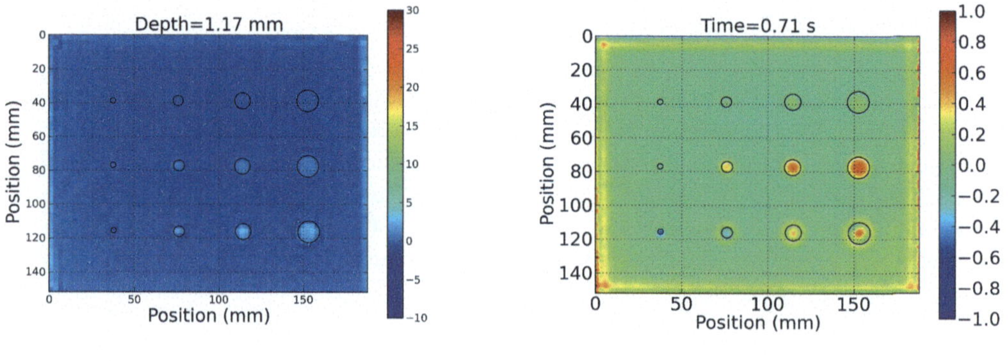

(a) (b)

Fig. 8.1 (**a**) Model-based inversion results from a flat-bottom hole sample at a depth of 1.17 mm, (**b**) Thermographic signal reconstruction second derivative at corresponding time of 0.71 s

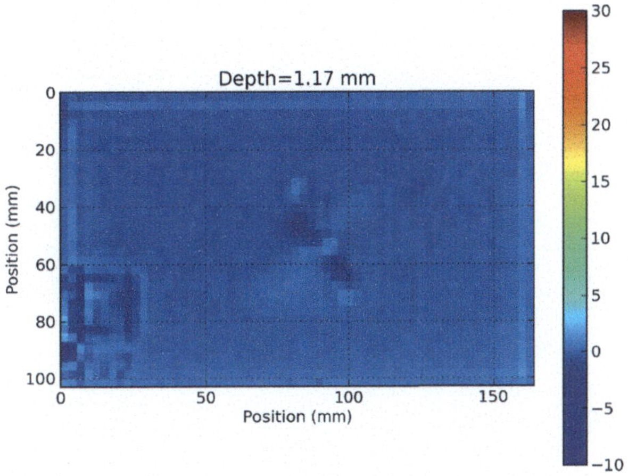

Fig. 8.2 Model-based inversion results for 1.17 mm depth from an impact-damaged carbon fiber plate

Acknowledgements This material is based on work supported by NASA Early Stage Innovation, grant #NNX15AD75G.

References

1. Maldague, X.: Theory and practice of infrared technology for nondestructive testing. Wiley, New York (2001)
2. Shepard, S.M., Lhota, J.R., Rebadeux, B.A., Wang, D., Ahmed, T.: Reconstruction and enhancement of active thermographic image sequences. Opt. Eng. **42**(5), 1337–1342 (2003)
3. Holland, S.D., Gregory, E., Schiefelbein, B.: Model-based inversion of flash thermography nondestructive evaluation measurements of composites. In: Proceedings of the American Society Composites, Williamsburg (2016)

Chapter 9
Experimentally Enhanced Computations: Calibration Methodology for an Anisotropic Metal, Part I: Traditional Approach

Edmundo Corona, Sharlotte Kramer, and Amanda Jones

Abstract The Experimentally Enhanced Computations project was motivated by the combined availability of advanced diagnostics such as digital image correlation (DIC) and non-quadratic, anisotropic yield functions for metals that have been implemented in computational mechanics codes. Here we propose to investigate the use of DIC combined with inverse methods as an alternative to traditional model calibration methods. The objective of this novel approach is to reduce the number of tests required for calibration, thus expediting the calibration process.

Keywords Model calibration · Computational modeling · Anisotropy · Digital image correlation · Aluminum

In Part I, we discuss the traditional calibration approach as the first installment of the investigation. In brief, traditional calibration procedures utilize simple specimen geometries under homogenous states of stress. To provide a benchmark for our proposed calibration technique, we explore calibration of the material in a tubular aluminum 7079 extrusion using tensile specimens cut along 12 orientations in the three planes as shown in Fig. 9.1.

The tensile specimens had square cross-sections in order to conveniently estimate their Lankford ratios. The thickness of the extrusion wall was large enough to cut specimens in the through-thickness direction. This allowed calibrations that were based exclusively on experimental results. Figure 9.2a shows the set-up used to test specimens cut in plane A while Fig. 9.2b shows the same for specimens in planes B and C. DIC was used in both cases to measure strains on two perpendicular faces of the specimens. For the large specimens, strain gages were used on the sides that were not imaged to complement the DIC measurements.

Measurements of yield stress and Lankford ratios in 12 uniaxial specimens were used to calibrate two anisotropic yield functions: Hill [1] with six anisotropy parameters and Barlat et al. [2] with 18 parameters. The calibrations were conducted by minimizing the difference between the yield stresses and Lankford ratios calculated from both the yield functions and experimental data following Tardif and Kyriakides [3]. The results depended to some degree on choices that must be made during calibration. In general, the results show that Barlat's yield function, with its larger number of parameters, can fit the measured material response significantly better.

Figure 9.3 shows the results of the anisotropy calibrations for the yield function from Barlat et al. Figure 9.3a shows the target value of the yield function (Y) against the values predicted by the yield function for the 12 specimen orientations (open points). Figure 9.3b shows a similar comparison of the target Lankford ratios and those predicted by the yield function. The comparisons are reasonably accurate in all cases, but still require a significant number of tests to fully calibrate the system. The results of the calibration for Hill's function showed more scatter, because of the smaller number of anisotropy parameters, but similar trends.

E. Corona · S. Kramer · A. Jones (✉)
Sandia National Laboratories, Albuquerque, NM, USA
e-mail: ajones1@sandia.gov

A. Baldi et al. (eds.), *Residual Stress, Thermomechanics & Infrared Imaging, Hybrid Techniques and Inverse Problems, Volume 7*, Conference Proceedings of the Society for Experimental Mechanics Series, https://doi.org/10.1007/978-3-319-95074-7_9

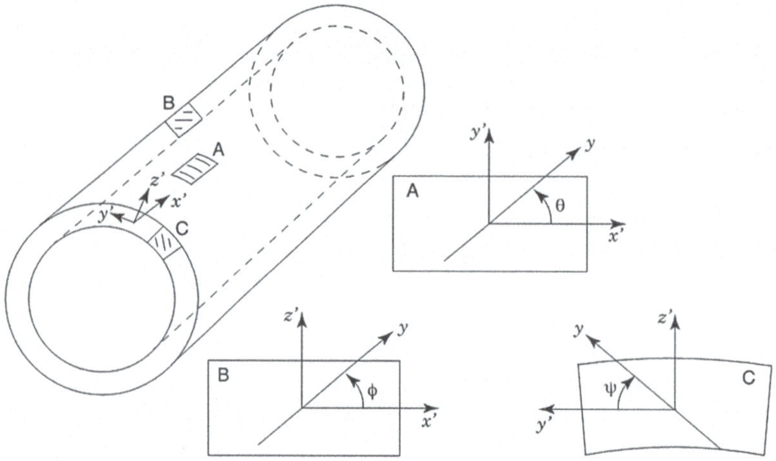

Fig. 9.1 Definition of planes and coordinates in cylindrical extrusion

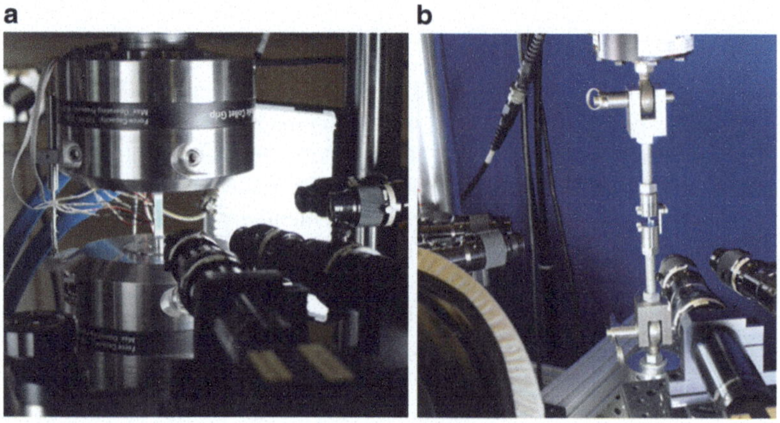

Fig. 9.2 Tension test set-ups. (**a**) In-plane specimens and (**b**) through-thickness specimens

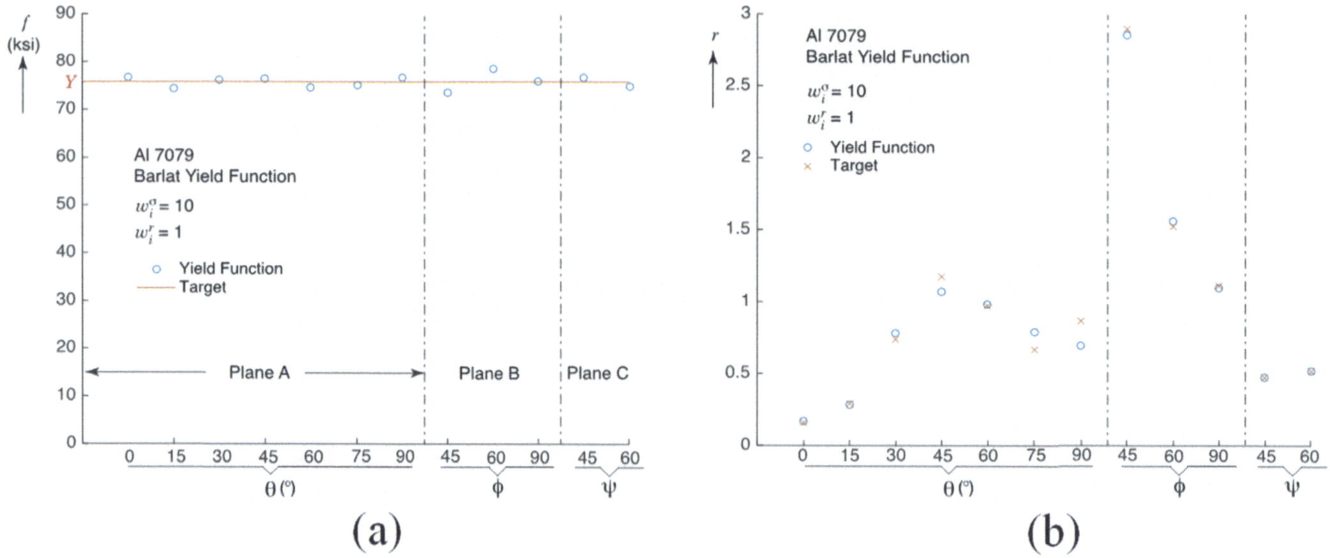

Fig. 9.3 Calibration results for a Barlat yield function. (**a**) Yield stress and (**b**) Lankford ratios

Acknowledgements Sandia National Laboratories is a multimission laboratory managed and operated by National Technology and Engineering Solutions of Sandia, LLC., a wholly owned subsidiary of Honeywell International, Inc., for the U.S. Department of Energy's National Nuclear Security Administration under contract DE-NA-0003525.

References

1. Hill, R.: A theory of the yielding and plastic flow of anisotropic materials. Proc. R. Soc. Lond. A Math. Phys. Sci. **193**, 281–297 (1948)
2. Barlat, F., Aretz, H., Yoon, J.W., Karabin, M.E., Brem, J.C., Dick, R.E.: Linear transformation-based anisotropic yield functions. Int. J. Plast. **21**, 1009–1039 (2005)
3. Tardif, N., Kyriakides, S.: Determination of anisotropy and material hardening for aluminum sheet metal. Int. J. Solids Struct. **49**, 3496–3506 (2012)

Chapter 10
Mechanical Response and Energy Stored During Deformation of Crystallizing TPU

Abdelmonem Lachhab, Eric Robin, Jean-Benoît Le Cam, Frédéric Mortier, Yvon Tirel, and Frédéric Canévet

Abstract The present study investigates the thermomechanical behavior of closed-cell TPU foams. The effects of the density and the loading conditions on the softening, the residual strain and the hysteresis have first been characterized. The thermal responses exhibit numerous particularities. First, a threshold effect in terms of the density on the self-heating has been highlighted. Second, entropic effects are strongly weighted by energetic effects (internal energy variations) during the deformation. Typical changes in the thermal response highlight that SIC and crystallite melting occur during the deformation. The characteristic stretches of this phenomenon evolve with the maximum stretch applied. The lower the density, the lower the crystallinity. In the second part of this study, a complete energy balance is carried out during cyclic deformation of compact and foamed crystallizing TPUs. Results show that viscosity is not the only phenomenon involved in the hysteresis loop formation: a significant part of the mechanical energy brought is not dissipated into heat and is stored by the material when the material changes its microstructure, typically when it is crystallizing. Some of this energy is released during unloading, when melting occurs, but with a different rate, which contributes to the hysteresis loop. The part of the mechanical energy stored by the material has been quantified to investigate the effects of the loading rate and the void volume fraction on the energetic response of TPU. These effects cannot be predicted from the mechanical responses and the present study provides therefore information of importance to better understand and model the effects of the density and the loading conditions on the thermomechanical behavior of closed-cell TPU foams.

Keywords TPU foam · Intrinsic dissipation · Energy stored · Strain-induced crystallization · Infrared thermography

10.1 Introduction

The thermoplastic polyurethanes (TPUs) are subjected to many anelastic phenomena such as mechanical hysteresis, residual stretch and softening [1–6], which strongly depend on the material formulation. When foamed, TPU becomes more deformable and are still dissipative [7–10]. The energy dissipation capacity of TPU is classically measured as the mechanical hysteresis area, but no study investigates the physical origin of this dissipation, while three phenomena could be responsible for this mechanical hysteresis: the intrinsic dissipation (d_1), due to internal friction and/or damage, which leads to self-heating; the thermal dissipation (d_2) and the part of the mechanical energy used by the material to change its microstructure ($W_{structure}$), see for instance recent studies on polyurea [11], natural rubber [12] and nitrile rubber [13].

Such analysis is classically carried out in metallic materials [14–17]. Polymers have then benefited from this approach [18–21]. Concerning elastomers, only two recent studies by Le Cam and co-workers investigate the energetic behavior and the energy storage during deformation [12, 13]. To further discuss on the relative contribution of the energy stored in the hysteresis loop of rubbers, they proposed a ratio γ_{se}, written in terms of energies over one mechanical cycle as follows:

A. Lachhab
Univ Rennes, CNRS, IPR (Institute de Physique de Rennes) – UMR 6251, Rennes, France

LC-DRIME, Joint Research Laboratory, Cooper Standard – Institut de Physique UMR 6251, Rennes Cedex, France

Cooper Standard France, Rennes, France

E. Robin · J. -B. Le Cam (✉)
Univ Rennes, CNRS, IPR (Institute de Physique de Rennes) – UMR 6251, Rennes, France

LC-DRIME, Joint Research Laboratory, Cooper Standard – Institut de Physique UMR 6251, Rennes Cedex, France
e-mail: jean-benoit.lecam@univ-rennes1.fr

F. Mortier · Y. Tirel · F. Canévet
Cooper Standard France, Rennes, France

A. Baldi et al. (eds.), *Residual Stress, Thermomechanics & Infrared Imaging, Hybrid Techniques and Inverse Problems, Volume 7*, Conference Proceedings of the Society for Experimental Mechanics Series, https://doi.org/10.1007/978-3-319-95074-7_10

$$\gamma_{se} = \frac{W_{stored}^{cycle}}{W_{hyst}^{cycle}} \tag{10.1}$$

W_{stored}^{cycle} is the energy stored over one cycle, W_{hyst}^{cycle} is the energy of the hysteresis loop.

The physical meaning of this ratio is different from the Taylor-Quinney coefficient, since it is not measured from a latent energy remaining after cold working: the energy stored in the elastomer network can be released. In the present work, a complete energy balance is carried out in order to determine the effects of the loading conditions and the void volume fraction on the energetic behavior of compact and foamed TPUs. Section 10.2 presents the thermodynamic framework. Section 10.3 sums up the technique used to elaborate the samples, the loading conditions and the thermal measurement during the mechanical tests. Section 10.4 gives the results. Concluding remarks close the paper.

10.2 Thermodynamical Framework

10.2.1 Total Strain Energy Density and Hysteresis Loop

The strain energy density W_{strain} (in J/m^3) is the energy brought mechanically to deform the material:

$$W_{strain}^{load} = \int_{load} \pi d\lambda \quad \text{and} \quad W_{strain}^{unload} = \int_{unload} \pi d\lambda \tag{10.2}$$

λ is the stretch defined as the ratio of the current length over the initial one. π is the nominal stress, defined as the force per unit of initial (undeformed) surface. If a hysteresis loop forms, the mechanical energy dissipated over one cycle W_{hyst}^{cycle} is defined as follows:

$$W_{hyst}^{cycle} = W_{strain}^{load} - W_{strain}^{unload} \tag{10.3}$$

From this energy, a quantity P_{hyst}^{cycle} is calculated in W/m^3. It is obtained by dividing W_{hyst}^{cycle} by the cycle duration.

10.2.2 Heat Power Densities

The homogeneous heat power density field s obtained during uniaxial tensile loadings writes as follows [22]:

$$s = \rho C \left(\dot{\theta} + \frac{\theta}{\tau} \right) \tag{10.4}$$

where θ denotes the temperature variation ($T - T_{ref}$) with respect to the equilibrium temperature T_{ref} in the undeformed state. In the present case, T_{ref} is constant and equal to the ambient temperature, which was constant during the tests. τ can be easily identified from a natural return to room temperature. Further details are provided in [23]. The heat power density can be divided into two terms that differ in nature: the intrinsic dissipation d_1 and the thermomechanical couplings stmc. Their integration with respect to time over one cycle is null. Therefore, the temporal integration of Eq. 10.4 over one mechanical cycle provides the energy density due to intrinsic dissipation $W_{intrinsic}^{cycle}$:

$$W_{intrinsic}^{cycle} = \int_{cycle} (s_{tmc} + d_1) = \int_{cycle} d_1 dt$$

$W_{intrinsic}^{cycle}$ is divided by the time duration to obtain the energy rate P_{d_1} due to intrinsic dissipation.

10.2.3 Energy Balance

The energy balance consists in comparing the energy contained in the mechanical hysteresis W_{hyst}^{cycle}, more precisely P_{hyst}^{cycle}, with P_{d_1}. If the both quantities are equal for a given cycle, all the mechanical energy of the hysteresis loop is converted into heat. Else, it is necessary to distinguish the part of the mechanical energy converted into heat from the one stored by the material to change its microstructure. For that purpose, the ratio γ_{se} recently proposed by Le Cam and co-workers [13] is used.

10.3 Experiments

The material considered here is a TPU referred to as Irogran R A87H4615 TPU marketed by the Huntsman corporation (The Woodlands, Texas, USA). The TPU is mixed with carbon dioxide in its supercritical state during injection, in order to form cells and to significantly decrease its density. The forming process is more precisely described in [6, 24]. The samples were obtained by cutting a bar with a water jet cutting device. Three densities were compared: 1.2, 0.7 and 0.5 kg/dm^3, the corresponding samples are denoted respectively $d_{1.2}$, $d_{0.7}$ and $d_{0.5}$ in the following. The density equal to 1.2 corresponds to the compact form. The sample geometry and the loading condition are given in Fig. 10.1.

Uniaxial tensile loadings were carried out with a 5543 Instron testing machine. Tests were performed under prescribed displacement. The signal shape was triangular to ensure a constant strain rate during loading and unloading.

10.3.1 Thermal Measurement

Temperature field measurements were performed using a FLIR X6540sc infrared camera. The acquisition frequency was set at 50 Hz. The thermal resolution (NETD) was equal to 20 mK at 25 °C. The calibration of the camera detectors was performed with a black body using a Non-Uniformity Correction (NUC) procedure. Temperature measurement is performed at the specimen centre, by averaging the temperature in a small zone of 5 × 5 pixels at the centre of the sample. As the zone moves during the test, a suitable movement compensation technique enables us to track this zone during the test (see [25–27] for further information on this technique).

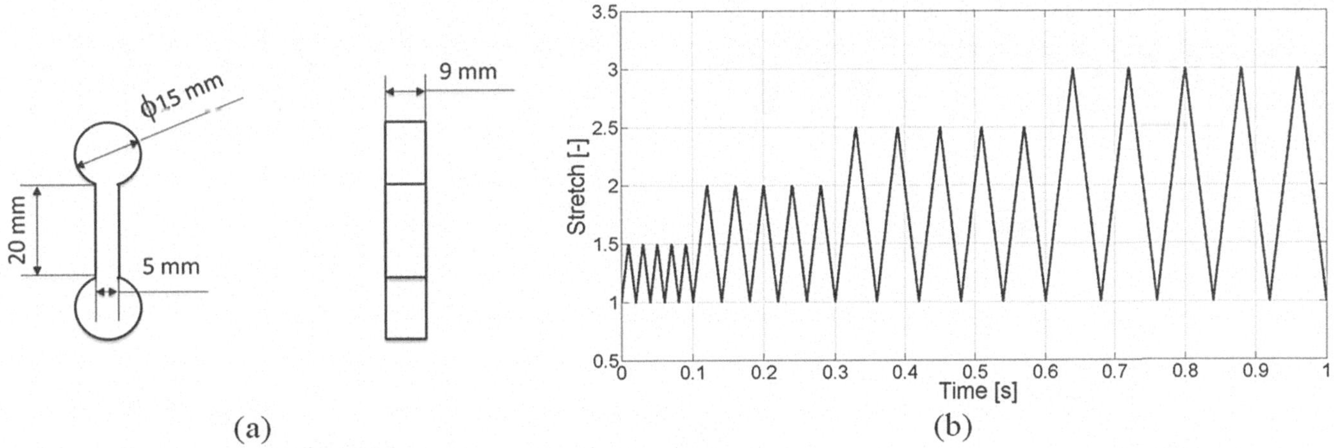

Fig. 10.1 (**a**) Sample geometry (**b**) loading conditions

10.4 Results and Discussion

10.4.1 Mechanical Responses

The mechanical responses obtained are precisely detailed in [6] and is therefore not recalled here. The main results can be summed up as follows:

- decreasing the density leads to a decrease in the global stiffness and in the hysteresis area, slightly increases the residual stretch and does not affect significantly the stress softening;
- the loading rate has no significant influence on the stress softening, while it influences more the residual stretch and the hysteresis area. It should be noted that the loading rate does not affect significantly the stiffness, which is a less intuitive result that will be discussed in the following.

10.4.2 Calorimetric Responses

Parameter τ has been characterized for each test configuration and material tested and the corresponding heat power densities have been determined by applying Eq. 10.4. Results obtained are presented in Fig. 10.2a, b for the three densities tested at ±100 mm/min and ±300 mm/min, respectively.

The heat power density increases when the TPU is stretched and decreases when the stress is released. The higher the density, the higher the maximum heat power densities produced. The material accommodates each time the maximum stretch is increased. Let us now consider the calorimetric response versus the stretch for the 3 densities at ±300 mm/min, presented in Fig. 10.3. TPU $d_{1.2}$ is chosen as the reference one. For the first two sets of stretches applied ($\lambda = 1.5$ and $\lambda = 2$), heat power densities are symmetrical with respect to the abscissa axis. Therefore, no significant intrinsic dissipation is produced. For sets at a maximum stretch superior to 2, a strong change in the curve slope is observed during the loading (at about $\lambda = 2.1$ for the third set at $\lambda = 2.5$). As the mechanical response does not exhibit such a singularity, this cannot be due to the effect of the entropic coupling. This strong increase in the heat production is analogue to what is observed in NR when crystallizes [12,

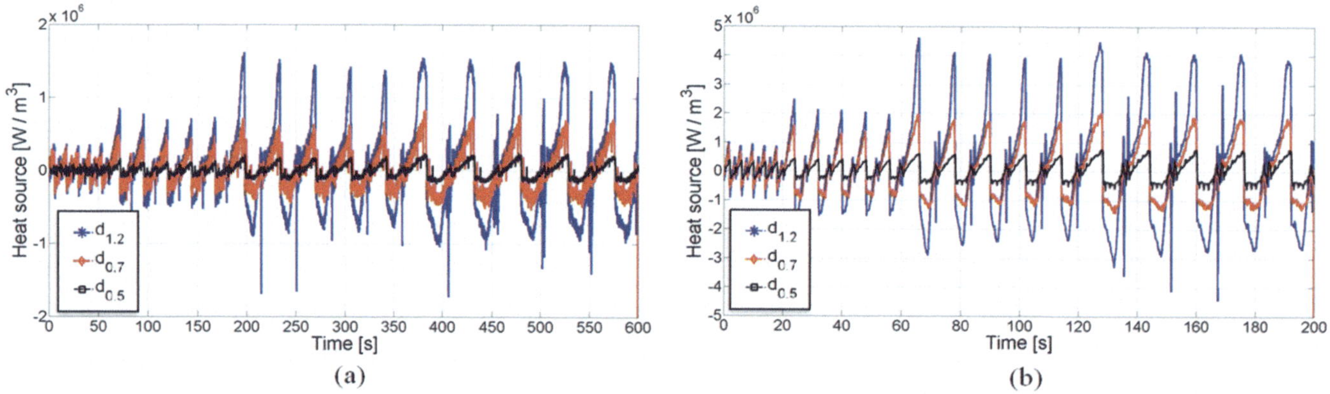

Fig. 10.2 Heat power density for the three densities. (a) ±100 mm/min. (b) ±300 mm/min

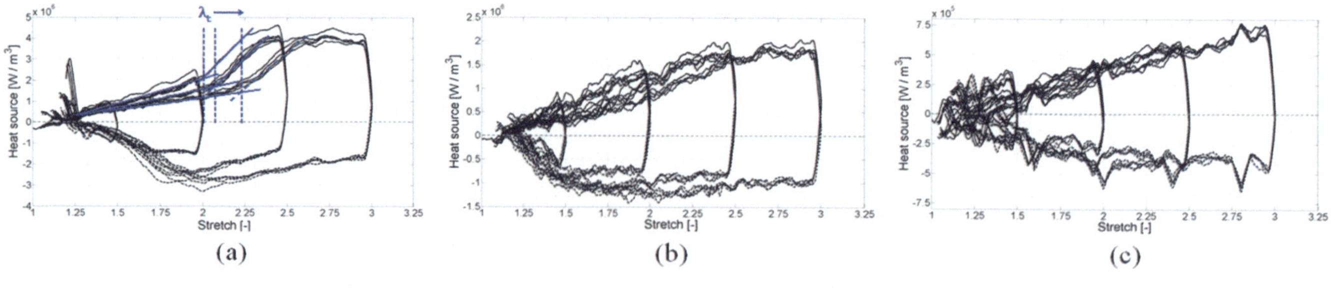

Fig. 10.3 Heat power density at ±300 mm/min for the three densities. (a) $d_{1.2}$. (b) $d_{0.7}$. (c) $d_{0.5}$

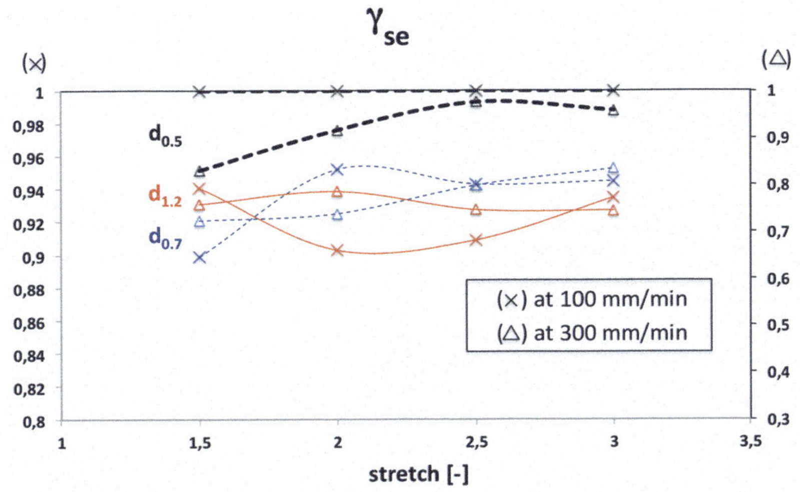

Fig. 10.4 γ_{se} in relation to the density, the loading rate and the stretch

23] and highlights that the present TPU is crystallizing under strain. The higher the maximum stretch applied, the higher the stretch at which the strong increase in the heat production is observed, and the higher the residual stretch. During unloading, a strong dissymmetry comparable to what is obtained in NR, is observed and strong heat absorption is observed at a stretch inferior to that at which SIC starts. The effects of the density and the loading rate will be more precisely detailed during the presentation.

10.4.3 Energy Balance

To further discuss on the energetic behavior and especially on physical origin of the mechanical hysteresis, Fig. 10.4 gives γ_{se} in relation to the density, the loading rate and the stretch. It should be noted that the calorific response is not stabilized from the second cycle, especially for results obtained samples $d_{0.5}$ at ±300 mm/min. In this case, P_{d_1} tends to zero, which means that γ_{se} would be close to 1 whatever the stretch applied. Such a representation opens a new way to investigate the thermomechanical and energetic behaviour of elastomers.

10.5 Conclusion

The present study investigates the thermomechanical and calorific behavior of TPU foams. Complete energy balances have been carried out in order to highlight the physical origin of the hysteresis loop. Different densities were tested under cyclic uniaxial tension at different loading rates and sample densities, including the compact state, which ensures a relevant characterization of the effect of the cell volume fraction on the TPU's formulation used. The calorimetric response has shown that the material's elasticity is mainly entropic, but it uses mechanical energy brought to it to change its microstructure. This is in good agreement with measurements recently performed in a compact TPU in [11]. For samples of the highest density, viscous effects are found to increase with the loading rate, the density and the maximum stretch applied. This is not observed for sample of the lowest density, for which the intrinsic dissipation remains low. For the lowest density, no self-heating was observed, even when the loading rate was increased by a factor 3. The fact that the material is able to store part of the mechanical energy probably explains why it is among the most resistant materials to the crack growth. Such a result is of importance when designing foamed TPU parts submitted to cyclic loadings.

Acknowledgements The authors thank the Cooper Standard France company for supporting this work and for fruitful discussions. The authors thank also the National Center for Scientific Research (MRCT-CNRS and MI-CNRS) and Rennes Metropole for supporting this work financially.

References

1. Blundell, D., Eeckhaut, G., Fuller, W., Mahendrasingam, A., Martin, C.: Real time SAXS/stress strain studies of thermoplastic polyurethanes at large strains. Polymer. **43**, 5197–5207 (2002)
2. Yeh, F., Hsiao, B., Sauer, B., Michael, S., Siesler, H.: In-situ studies of structure development during deformation of a segmented poly(urethane-urea) elastomer. Macromolecules. **36**, 1940–1954 (2003)
3. Unsal, E., Yalcin, B., Yilgor, I., Yilgor, E., Cakmak, M.: Real time mechano-optical study on deformation behavior of ptmo/chdi-based polyetherurethanes under uniaxial extension. Polymer. **50**, 4644–4655 (2009)
4. Bartolome, L., Aurrekoetxea, J., Urchegui, M.A., Tato, W.: The influences of deformation state and experimental conditions on inelastic behaviour of an extruded thermoplastic polyurethane elastomer. Mater. Des. **49**, 974–980 (2013)
5. Qi, H., Boyce, M.: Stress-strain behavior of thermoplastic polyurethanes. Mech. Mater. **37**, 81–839 (2005)
6. Lachhab, A., Robin, E., Le Cam, J.-B., Mortier, F., Tirel, Y., Canevet, F.: Thermomechanical analysis of closed-cell polymeric foams subjected to cyclic loading: anelasticity, self-heating and stressinduced crystallization. Polymer. **126**, 19–28 (2017)
7. Blatz, P.J., Ko, W.L.: Application of finite elastic theory to the deformation of rubbery materials. Trans. Soc. Rheol. **6**, 223–252 (1962)
8. Saint-Michel, F., Chazeau, L., Cavaillé, J.-Y., Chabert, E.: Mechanical properties of high density polyurethane foams: I. Effect of the density. Compos. Sci. Technol. **66**, 2700–2708 (2006)
9. Alzoubi, M.F., Tanbour, E.Y., Al-Waked, R.: Compression and hysteresis curves of nonlinear polyurethane foams under different densities, strain rates and different environmental conditions. ASME. **9**, 101–109 (2011)
10. Ugarte, L., Saralegi, A., Fernandez, R., Martin, L., Corcuera, M.A., Eceiza, A.: Flexible polyurethane foams based on 100% renewably sourced polyols. Ind. Crop. Prod. **62**, 545–551 (2014)
11. Mott, P., Giller, C., Fragiadakis, D., Rosenberg, D., Roland, C.: Deformation of polyurea: where does the energy go? Polymer. **105**, 227–233 (2016)
12. Le Cam, J.-B.: Energy storage due to strain-induced crystallization in natural rubber: the physical origin of the mechanical hysteresis. Polymer. **127**, 166–173 (2017)
13. Loukil, M., Corvec, G., Robin, E., Miroir, M., Le Cam, J.-B., Garnier, P.: Stored energy accompanying cyclic deformation of filled rubber. Eur. Polym. J. **98**, 448–455 (2018)
14. Chrysochoos, A.: Energy balance for elastic plastic deformation at finite strain (in french). Journal de Mecanique theorique et appliquee. **5**, 589–614 (1985)
15. Mason, J., Rosakis, A., Ravichandran, G.: On the strain and strain rate dependence of the fraction of plastic work converted to heat: an experimental study using high speed infrared detectors and the kolsky bar. Mech. Mater. **17**, 135–145 (1994)
16. Rittel, D.: On the conversion of plastic work to heat during high strain rate deformation of glassy polymers. Mech. Mater. **31**, 131–139 (1999)
17. Oliferuk, W., Maj, M., Raniecki, B.: Experimental analysis of energy storage rate components during tensile deformation of polycrystals. Mater. Sci. Eng. A. **374**, 77–81 (2004)
18. Rittel, D.: An investigation of the heat generated during cyclic loading of two glassy polymers. Part I: experimental. Mech. Mater. **32**, 131–147 (2000)
19. Rittel, D., Rabin, Y.: An investigation of the heat generated during cyclic loading of two glassy polymers. Part II: thermal analysis. Mech. Mater. **32**, 149–159 (2000)
20. Benaarbia, A., Chrysochoos, A., Robert, G.: Kinetics of stored and dissipated energies associated with cyclic loadings of dry polyamide 6.6 specimens. Polym. Test. **34**, 155–167 (2014)
21. Benaarbia, A., Chrysochoos, A., Robert, G.: Influence of relative humidity and loading frequency on the PA6.6 thermomechanical cyclic behavior: Part II. Energy aspects. Polym. Test. **41**, 92–98 (2015)
22. Chrysochoos, A.: Analyse du comportement des matériaux par thermographie infra rouge. Colloque Photomécanique. **95**, 201–211 (1995)
23. Samaca Martinez, J.R., Le Cam, J.-B., Balandraud, X., Toussaint, E., Caillard, J.: Mechanisms of deformation in crystallizable natural rubber. Part 2: quantitative calorimetric analysis. Polymer. **54**, 2727–2736 (2013)
24. Primel, A., Ferec, J., Ausias, G., Tirel, Y., Veille, J.-M., Grohens, Y.: Solubility and interfacial tension of thermoplastic polyurethane melt in super-critical carbon dioxide and nitrogen. J. Supercrit. Fluid. **122**, 52–57 (2017)
25. Pottier, T., Moutrille, M.-P., Le Cam, J.-B., Balandraud, X., Grédiac, M.: Study on the use of motion compensation technique to determine heat power densities. Application to large deformations on cracked rubber specimens. Exp. Mech. **49**, 561–574 (2009)
26. Toussaint, E., Balandraud, X., Le Cam, J.-B., Grediac, M.: Combining displacement, strain, temperature and heat power density fields to investigate the thermomechanical response of an elastomeric specimen subjected to large deformations. Polym. Test. **31**, 916–925 (2012)
27. Samaca Martinez, J.R., Le Cam, J.-B., Balandraud, X., Toussaint, E., Caillard, J.: Mechanisms of deformation in crystallizable natural rubber. Part 1: thermal characterization. Polymer. **54**, 2717–2726 (2013)

Chapter 11
Measuring Strain-Induced Crystallinity in Rubbers from IR Thermography

Jean-Benoît Le Cam

Abstract The crystallinity of stretched crystallizable rubbers is classically investigated using X-ray diffraction (XRD). In the present study, we propose a new method based on temperature measurement and quantitative calorimetry to determine rubber crystallinity during mechanical tests as those carried out with conventional mechanical testing machines. For that purpose, heat power density are first determined from temperature variation measurements and the heat diffusion equation. The increase in temperature due to strain-induced crystallization (SIC) is then deduced from the heat power density by subtracting the part due to elastic couplings. The heat capacity, the density and the enthalpy of fusion are finally used to calculate the crystallinity from the temperature variations due to SIC. The characterization of the stress-strain relationship is not required. Furthermore, nonentropic contributions to rubber elasticity are taken into account if any. This alternative crystallinity measurement method is a user-friendly measurement technique, which is well adapted to most of the mechanical tests. It opens numerous perspectives in terms of high speed and full crystallinity field measurements.

Keywords Crystallinity · Strain-induced crystallization · Natural rubber · Infrared thermography · Calorimetry

11.1 Introduction

The strain-induced crystallization (SIC) of polymer, and especially of rubber, is classically investigated by using X-ray diffraction (XRD) [1]. Such measurement method provides information of importance on the crystalline phase structure [2–5], the chain orientation [6] and the kinetics of crystallization [7, 8]. To measure crystallinity, XRD requires specific equipment and tests conditions. Moreover, quantifying small crystallinity changes is difficult below 2% [8]. To finish, in case of heterogeneous fields, each point of the crystallinity field is obtained successively, by scanning the zone under study [9–11] and the measurement zone is sometimes too large (for instance $300 \times 200\ \mu m^2$ in [10]) compared to the crystallinity gradient, typically at the crack tip of rubbers.

Developing an alternative measurement method could therefore be beneficial in many ways, typically for obtaining the crystallinity field instantaneously, at higher resolutions (temporal, spatial and on the value of the crystallinity) and with a conventional testing machine.

As SIC is well-known to be strongly exothermal, temperature or calorimetry measurements should be therefore a relevant alternative to determine the crystallinity. Indeed, in the 1970s Göritz and co-workers showed that crystallinity could be quantified accurately during stretching through calorimetric measurements [12]. Note that Boonstra was the first who suggested that crystallinity could be determined from the induced change in internal energy and the heat of fusion per repeating unit [13]. Using calorimetry is in the same line, because change in the internal energy can be taken into account with a "stretch calorimeter." Despite this, calorimetry under stretching was no longer used to measure crystallinity. Note that it has been nevertheless used to characterize the thermomechanical response and the intrinsic dissipation associated to the cyclic deformation of rubber [14–16].

Recently, Spratte et al. have taken up the idea that crystallinity χ can be measured from the thermal activity at the surface of rubber specimens [17]. Crystallinity was determined from the temperature change due to SIC T_{cryst} by considering that the crystallization energy of natural rubber can be approximated by the enthalpy of fusion ΔH:

J. -B. Le Cam (✉)
Univ Rennes, CNRS, IPR (Institute de Physique de Rennes) – UMR 6251, Rennes, France

LC-DRIME, Joint Research Laboratory, Cooper Standard – Institut de Physique UMR 6251, Rennes Cedex, France
e-mail: jean-benoit.lecam@univ-rennes1.fr

A. Baldi et al. (eds.), *Residual Stress, Thermomechanics & Infrared Imaging, Hybrid Techniques and Inverse Problems, Volume 7*, Conference Proceedings of the Society for Experimental Mechanics Series, https://doi.org/10.1007/978-3-319-95074-7_11

$$\chi(t) = \frac{\rho C T_{cryst}(t)}{\Delta H} \tag{11.1}$$

ρ and C are the density and the heat capacity, respectively. They are assumed to be independent of strain and temperature.

In this approach, only two contributions to temperature changes were distinguished by the authors: crystallinity change and entropy change. Energetic effects were therefore neglected [18], while they can influence significantly the thermal response [19–25].

In the present study, crystallinity is determined from heat power density instead of temperature, by using infrared thermography during the mechanical tests. In practice, any temperature measurement technique such as a pyrometer can be used. This method takes intrinsically into account both the non-entropic (energetic) and the entropic contributions to rubber elasticity. Moreover, the measurement of the stress is not required. As this technique provides a full thermal field, crystallinity heterogeneity can be characterized, at several thousand frames per second, each value of the field being obtained at the same time. The methodology used to determine the crystallinity during rubber deformation is first presented. Then, it is applied to experimental data available in the literature for unfilled NR. Concluding remarks close the paper.

11.2 Methodology for Measuring Crystallinity

11.2.1 Thermodynamical Framework

Most of mechanical tests are conducted under non-adiabatic conditions. The temperature measured is therefore affected by heat diffusion during the tests, possible temperature gradients at the surface of the undeformed specimen and external heat sources (for instance radiations). Therefore, changes in temperature are not only due to the material deformation, and the heat diffusion equation is used to determine the corresponding heat power density from temperature measurement. This quantity is intrinsic to the material deformation. After some calculations that are not detailed here, the heat diffusion equation can be rewritten in case of homogeneous heat power density field [26]:

$$s = \rho C \left(\dot{\theta} + \frac{\theta}{\tau} \right) \tag{11.2}$$

where θ denotes the temperature variation ($T - T_{ref}$) with respect to the equilibrium temperature T_{ref} in the undeformed state. In the present case, T_{ref} is constant and equal to the ambient temperature. In case where changes in ambient temperature occur, T_{ref} has to be corrected accordingly. τ is a parameter characterizing the heat exchanges between the specimen and its surroundings. It can be easily identified from a natural return to room temperature after a heating (or a cooling) for each testing configuration (machine used, environment, stretch level, etc). Further details are provided in [27].

11.2.2 Determination of the Crystallinity

The methodology proposed is composed of 4 steps presented in the diagram in Fig. 11.1. Input data required are the temperature variation θ, parameter τ and the thermo-physical parameters ρ, C and ΔH. The first step consists in calculating the heat power density by using Eq. 11.2. The typical calorific response of unfilled NR is illustrated by curve A. It exhibits a strong increase in the heat power density when crystallization starts. λ_c and λ_m are the stretches at which SIC starts and melting is complete, respectively. Step 2 consists in predicting the heat power density due to elastic couplings only by using a polynomial form ($C_1(I_1 - 3) + C_2(I_1 - 3)^2 + C_3(I_1 - 3)^3$). The thermal energy due to SIC is indeed the area located between the heat power density (curve A) and the part of the heat power density due to elastic couplings (curve B). The latter is not known once crystallization starts, but is known before. In the present study, we assume that the elastic couplings evolve in a similar way after crystallization starts. C_1, C_2 and C_3 are identified from the experimental heat power density before SIC starts. The heat power density due to SIC (A-B) is then computed in Step 3 for determining the temperature variation due to SIC T_{cryst}. Here, the numerical scheme is not centered, another numerical scheme can be used. The last step consists in calculating the crystallinity by applying Eq. 11.1. This method is very simple and does not require measuring the nominal stress variations nor characterizing possible non-entropic effects.

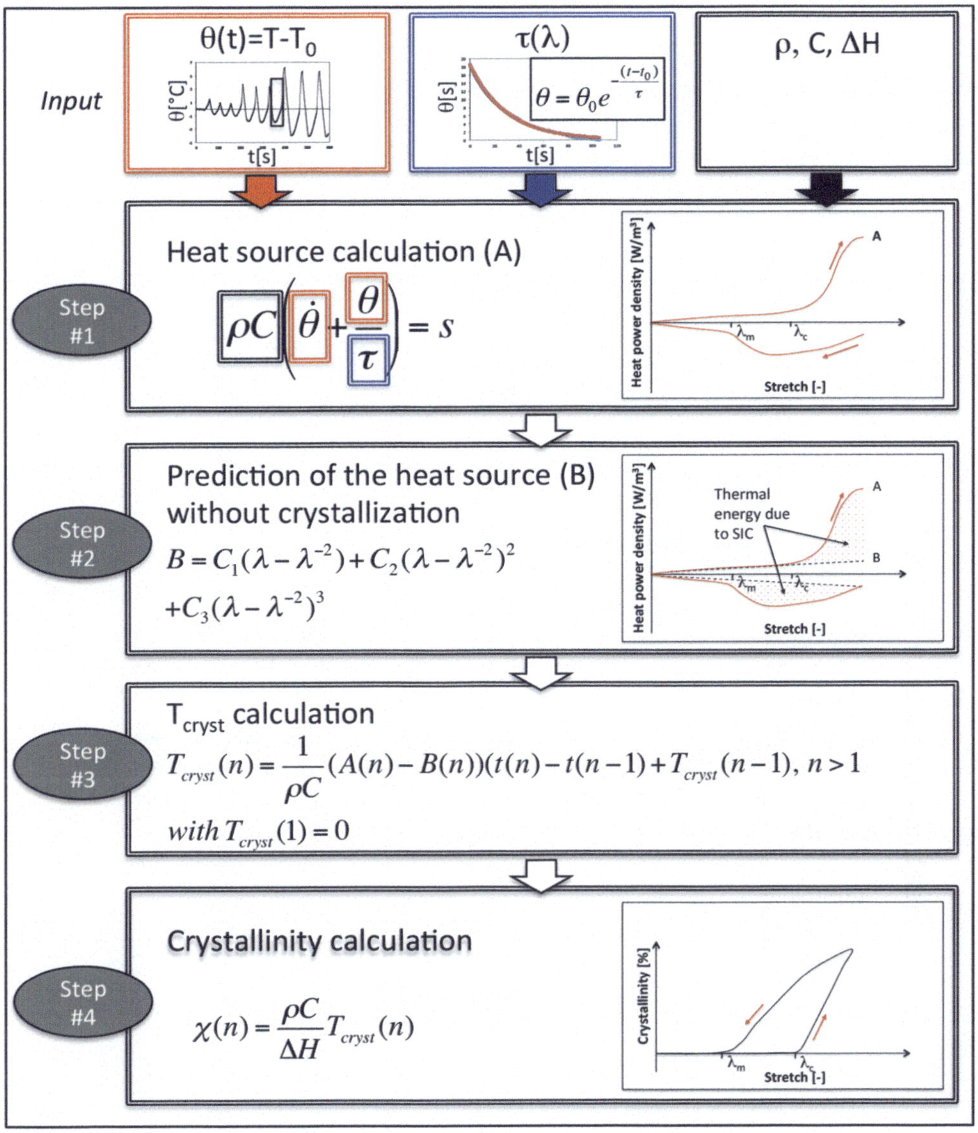

Fig. 11.1 Methodology for determining crystallinity

11.3 Application to an Unfilled Natural Rubber

The methodology proposed for measuring crystallinity is now applied to experimental data issued from the literature [27, 28], obtained with an unfilled natural rubber submitted to cyclic uniaxial tensile loading. The mechanical loading applied corresponded to four sets of three cycles at a maximum stretch of 2, 5, 6 and 7.5 at ±100 mm/min. It is presented in Fig. 11.2.

The stretch is defined as the ratio of the current length over the initial length. ρ and C are equal to 936 g/dm^3 and 1768 J/(kg · K). τ depends on the stretch and is defined as $\tau(\lambda) = 40.48 - 3.25\lambda$. Figure 11.3 depicts the heat power density in relation to the stretch. The heat power density is determined from the temperature measurements (Step #1) and appears in red color. It is due to elastic couplings and SIC effects. During loading (unloading), the heat power density is positive (negative) and increases (decreases) with the stretch. Once SIC starts, the heat power density evolution for loading and unloading are no longer symmetrical. During loading, the heat power density evolves in a quasi-linear manner until a stretch close to 4 is reached. A dissymmetry is observed for stretches higher than 4; the stretch level at which SIC starts. During unloading, the heat power density rate first increases in absolute value until reaching a stretch equal to 4, is constant until a stretch equal to 3.5 is reached, and then decreases.

The heat power density due to elastic couplings only, i.e. without accounting for SIC effects, is predicted with a polynomial form (Step #2) and appears in black color. The thermal energy due to SIC is the area located between the heat power density

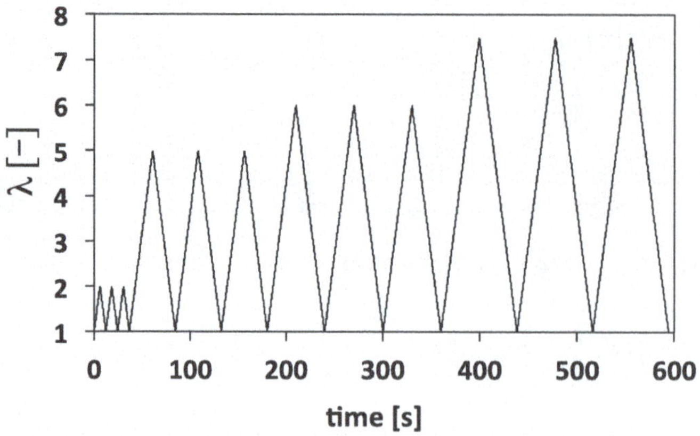

Fig. 11.2 Mechanical loading

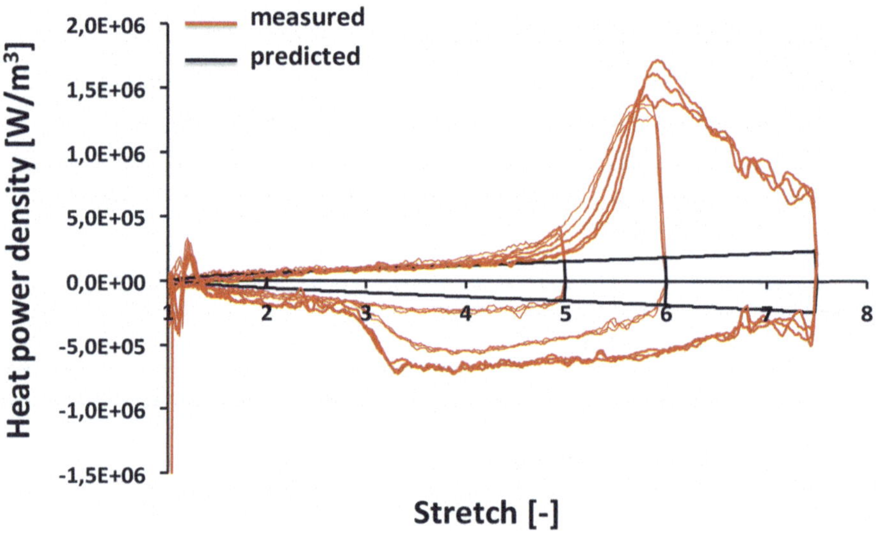

Fig. 11.3 Heat power density (in red) and predicted heat power density due to elastic couplings (in black) in relation to the stretch

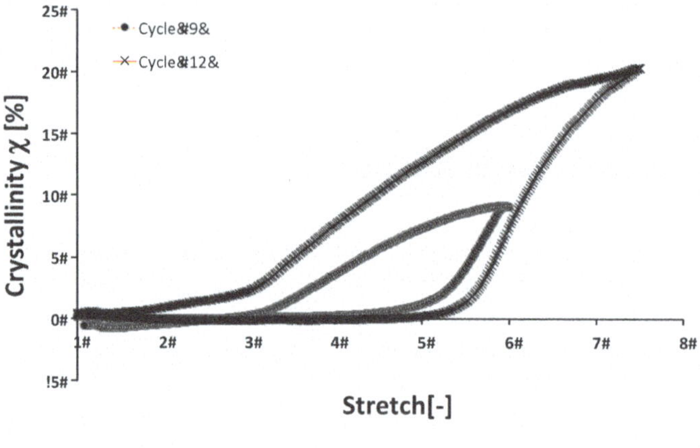

Fig. 11.4 Temperature change due to crystallinity in relation to the stretch (Step #4)

measured (curve A in Fig. 11.1) and the predicted part of the heat power density curve due to elastic couplings (curve B). The difference between the two curves is used to calculate T_{cryst} (Step #3).

The results being the same for cycles conducted at the same maximum stretch, only cycles #9 and #12 at respectively $\lambda = 6$ and $\lambda = 7.5$ are presented in Fig. 11.4. The crystallinity is obtained by dividing T_{cryst} by the enthalpy of crystallite fusion (59.9 J/cm^3, see [29]) and multiplying by the density and the heat capacity (Step #4). The crystallinity versus stretch is given in Fig. 11.3. The comparison with results obtained from XRD will be discussed during the presentation.

11.4 Conclusion

In this study, a new measurement technique is presented to determine the crystallinity in stretched crystallizing rubber using IR thermography. Temperature variation due to SIC is determined from the corresponding heat power density, i.e. due to SIC. Characterizing the stress-strain relationship as well as the non-entropic contributions to rubber elasticity is not required. It is therefore well adapted to characterize heterogeneous crystallinity field, i.e. where the stress-strain relationship is not known. Crystallinity obtained as a function of the stretch provides comparable results as those obtained by XRD measurements. The method proposed is therefore a realistic alternative, user-friendly and accurate, to measure crystallinity of stretched crystallizing rubbers from any temperature measurement technique and quantitative calorimetry. Furthermore, it is well adapted to most of the mechanical tests performed with conventional testing machine, which enables us to investigate the effect of SIC on fatigue resistance of rubber [15, 30, 31].

References

1. Katz, J.R.: Röntgenspektrographische Untersuchungen am gedehnten Kautschuk und ihre mögliche Bedeutung für das Problem der Dehnungseigenschaften dieser Substanz. Naturw. **4**, 410–416 (1925)
2. Bunn, C.W.: Molecular structure and rubber-like elasticity. I. The crystal structures of Gutta-Percha, rubber and polychloroprene. Proc. R. Soc. London, Ser. 1. **180**, 40–66 (1942)
3. Takahashi, Y., Kumano, T.: Crystal structure of natural rubber. Macromolecules. **37**, 4860 (2004)
4. Immirzi, A., Tedesco, C., Monaco, G., Tonelli, A.E.: Crystal structure and melting entropy of natural rubber. Macromolecules. **38**, 1223 (2005)
5. Rajkumar, G., Squire, J.M., Arnott, S.: A new structure for crystalline natural rubber. Macromolecules. **39**, 7004 (2006)
6. Toki, S., Sics, I., Ran, S.F., Liu, L.Z., Hsiao, B.S., Murakami, S., Tosaka, M., Kohjiya, S., Poompradub, S., Ikeda, Y., Tsou, A.H.: Strain-induced molecular orientation and crystallization in natural and synthetic rubbers under uniaxial deformation by in-situ synchrotron X-ray study. Rubber Chem. Technol. **77**, 317–335 (2004)
7. Toki, S., Fujimaki, T., Okuyama, M.: Strain-induced crystallization of natural rubber as detected real-time by wide-angle x-ray diffraction technique. Polymer. **41**, 5423–5429 (2000)
8. Trabelsi, S., Albouy, P.-A., Rault, J.: Effective local deformation in stretched filled rubber. Macromolecules. **36**, 9093–9099 (2003)
9. Brüning, K., Schneider, K., Roth, S.V., Heinrich, G.: Strain-induced crystallization around a crack tip in natural rubber under dynamic load. Polymer. **54**(22), 6200–6205 (2013)
10. Rublon, P., Huneau, B., Verron, E., Saintier, N., Beurrot, S., Leygue, A., Mocuta, C., Thiaudière, D., Berghezan, D.: Multiaxial deformation and strain induced crystallization around a fatigue crack in natural rubber. Eng. Fract. Mech. **123**, 59–69 (2014)
11. Demassieux, Q.: Structural changes in the process zone of a cyclic fatigue crack in filled natural rubber. PhD thesis, Université Pierre et Marie Curie (2016)
12. Göritz, D., Müller, F.H.: Die kalorimetrische Erfassung der Dehnungskristallisation Polymerer. Kolloid-Zeitschrift und Zeitschrift für Polymere. **241**(20), 1075–1079 (1970)
13. Boonstra, B.B.S.T.: Some properties of vulcanized rubber under strain – degree of crystallization as calculated from temperature coefficient of elastic tension. Ind. Eng. Chem. **4**(2), 3:362–3:365 (1951)
14. Samaca Martinez, J.R., Le Cam, J.B., Balandraud, X., Toussaint, E., Caillard, J.: New elements concerning the Mullins effect: a thermomechanical analysis. Eur. Polym. J. **55**, 98–107 (2014)
15. Samaca Martinez, J.R., Balandraud, X., Toussaint, E., Le Cam, J.B., Berghezan, D.: Thermomechanical analysis of the crack tip zone in stretched crystallizable natural rubber by using infrared thermography and digital image correlation. Polymer. **55**, 6345–6353 (2014)
16. Samaca Martinez, J.R., Balandraud, X., Toussaint, E., Le Cam, J.B., Berghezan, D.: Heat and strain measurements at the crack tip of filled rubber under cyclic loadings using full field techniques. Mech. Mater. **81**, 62–71 (2015)
17. Spratte, T., Plagge, J., Wunde, M., Klueppel, M.: Investigation of strain-induced crystallization of carbon black and silica filled natural rubber composites based on mechanical and temperature measurements. Polymer. **115**, 12–20 (2017)
18. Meyer, K.H., Ferri, C.: Sur l'élasticité du caoutchouc. Helv. Chim. Acta. **18**, 570–589 (1935)
19. Wolf, F.P., Allen, G.: The energetic contribution to rubber elasticity in the range of small uniaxial compression and moderate elongation. Polymer. **16**, 209–217 (1975)
20. Allen, G., Bianchi, U., Price, C.: Thermodynamics of elasticity of natural rubber. Trans. Faraday Soc. **59**, 2493 (1963)
21. Allen, G., Kirkham, M.J., Padget, J., Price, C.: Thermodynamics of rubber elasticity at constant volume. Trans. Faraday Soc. **67**, 1278 (1971)
22. Shen, M.C.: Internal energy contribution to the elasticity of natural rubber. Macromolecules. **2**, 358–364 (1969)
23. Le Cam, J.-B.: Energy storage due to strain-induced crystallization in natural rubber: the physical origin of the mechanical hysteresis. Polymer. **127**, 166–173 (2017)
24. Treloar, L.R.G.: The elasticity and related properties of rubbers. Rep. Prog. Phys. **36**(7), 755 (1973)
25. Heinrich, G., Kaliske, M., Klüppel, M., Mark, J.E., Straube, E., Vilgis, T.A.: The thermoelasticity of rubberlike materials and related constitutive laws. J. Macromol. Sci. Part A. **40**, 87–93 (2003)
26. Chrysochoos, A.: Analyse du comportement des matériaux par thermographie infra rouge. Colloque Photomécanique. **95**, 201–211 (1995)
27. Samaca Martinez, J.R., Le Cam, J.-B., Balandraud, X., Toussaint, E., Caillard, J.: Mechanisms of deformation in crystallizable natural rubber. Part 2: quantitative calorimetric analysis. Polymer. **54**, 2727–2736 (2013)

28. Samaca Martinez, J.R., Le Cam, J.-B., Balandraud, X., Toussaint, E., Caillard, J.: Mechanisms of deformation in crystallizable natural rubber. Part 1: thermal characterization. Polymer. **54**, 2717–2726 (2013)
29. Roberts, D.E., Mandelkern, L.: Thermodynamics of crystallization in high polymers. Natural rubber. Rubber Chem. Technol. **28**(3), 718–727 (1955)
30. Le Cam, J.-B., Huneau, B., Verron, E.: Fatigue damage in carbon black filled natural rubber under uni- and multiaxial loading conditions. Int. J. Fatigue. **52**, 82–94 (2013)
31. Le Cam, J.-B., Huneau, B., Verron, E., Gornet, L.: Description of fatigue damage in carbon black filled natural rubber. Fatigue Fract. Eng. Mater. Struct. **31**(12), 1031–1038 (2008)

Chapter 12
Mechanical and Thermomechanical Characterization of Different Leathers

Noëlie Di Cesare, Guillaume Corvec, Xavier Balandraud, Jean-Benoît Le Cam, and Julien Gauffreteau

Abstract Leather materials are able to undergo various strain and stress states during their elaboration process and their use in numerous applications. Although the experimental mechanical response in tension of leathers has been studied in the literature for decades, scarce information is available on the nature of their elasticity and more generally on their thermo-mechanical behaviors. In the present study, two leathers were tested under uniaxial cyclic loading while temperature changes were measured at the specimens' surface by infrared thermography. The heat power at the origin of the temperature changes was then determined by using an adequate version of heat diffusion equation which is applicable to homogeneous tests. Results enabled us to discuss on the physical nature of the thermoelastic coupling in leathers. Intrinsic dissipation caused by the mechanical irreversibility was also detected. Distinct behaviors are evidenced as a function of the type of leathers.

Keywords Infrared thermography · Leather · Heat source · Thermoelastic coupling · Mechanical dissipation

12.1 Introduction

Although the experimental mechanical response in tension of leathers has been studied in the literature for decades [1–5], scarce information is available on the nature of their elasticity and more generally on their thermomechanical behaviors. Some physical phenomena involved in their deformation mechanism have been proposed to model their mechanical response in terms of stiffness, rupture or relaxation. Beyond mechanical tests, some measurements carried out by using infrared (IR) thermography have been performed to analyze the temperature variations at the surface of leathers. Two types of approach are reported in the literature:

- application to non-destructive control of final leather products [6]. In this case, pulsed-phase thermography was performed to evidence hidden scratches and the texture in depth, and finally to evaluate the quality of the leather;
- analysis of the thermal response of leathers under mechanical loading [7–9]. More precisely, applications concerned the self-heating under cyclic loading, with the main objective of determining a limit of acceptable damage for leather products.

The present study belongs to the second type of approach and aims at studying the thermomechanical response of leathers under cyclic mechanical loading, more precisely under load-unload cycles. Due to thermomechanical couplings, any material produces or absorbs heat when stretched. In particular, thermoelastic couplings occur in any material. Two types of thermoelastic couplings can be distinguished: a "classical" thermoelastic coupling, which is sometimes named *isentropic coupling* [10] and occurs in metallic materials and polymers below their glass transition temperature; an *entropic coupling,* which occurs in rubber-like materials [11]. Rubber-like materials actually feature both types of thermoelastic coupling. Even though the entropic coupling is preponderant, the thermoelastic coupling occurs and is response for the

N. Di Cesare
Université Bretagne Sud, IUT de Lorient-Pontivy, IRDL, FRE CNRS 3744, Pontivy, France

G. Corvec · J.-B. Le Cam (✉)
Univ Rennes, CNRS, IPR (Institute de Physique de Rennes) – UMR 6251, Rennes, France

LC-DRIME, Joint Research Laboratory, Cooper Standard – Institut de Physique UMR 6251, Rennes Cedex, France
e-mail: jean-benoit.lecam@univ-rennes1.fr

X. Balandraud
Université Clermont Auvergne, CNRS, SIGMA Clermont, Institut Pascal, Clermont-Ferrand, France

J. Gauffreteau
SIGMA Clermont, Campus Design, Matériaux & Innovation, Aubière Cedex, France

A. Baldi et al. (eds.), *Residual Stress, Thermomechanics & Infrared Imaging, Hybrid Techniques and Inverse Problems, Volume 7*, Conference Proceedings of the Society for Experimental Mechanics Series, https://doi.org/10.1007/978-3-319-95074-7_12

thermoelastic inversion at low strains [12, 13]. Any mechanical irreversibility occurring in a material is also accompanied by heat production: plasticity, fatigue damage and viscosity are all associated with a heat production leading to material self-heating, i.e. a global increase in temperature. Measuring temperature changes at the surface of tested specimens is thus relevant to improve our understanding of materials' behavior. However, the analysis of the heat sources provides more relevant information than the analysis of the temperature changes (By "heat source", we mean the heat power density in W/m^3 produced or absorbed by the material itself due to a change in the mechanical state). Indeed, temperature is affected by the heat diffusion in the specimen and with the outside of the specimen (by conduction with the jaws of the testing machine, by convection with ambient air) and by radiations. Therefore, heat source is a "more intrinsic" quantity for investigating damage and deformation mechanisms. The heat source fields are reconstructed from the temperature fields and the heat diffusion equation [14, 15]. A zero-dimensional (0D) version of this equation can be employed when heat sources are homogeneous in the specimen, typically when the strain fields are homogeneous. This approach was used for rubbers [16–22], shape-memory alloys [23–29], polyamide 6.6 [30–34], PMMA [35], aluminum alloy [36], steel [37], copper [35, 38]. To the best knowledge of the authors, heat source calculation has never been performed in case of leathers.

In the present study, two leathers were tested under uniaxial cyclic loading while temperature changes were measured at the specimens' surface by using IR thermography. The heat power density at the origin of the temperature changes was then determined using the 0D version of the heat diffusion equation to analyze the thermomechanical response of the tested leathers.

12.2 Experimental Set-Up

The experimental set-up is presented in Fig. 12.1. Tests were conducted with a uniaxial testing machine featuring two actuators. Specimens were cut in a split cow leather (L1) and a grain calf leather (L2). A sample made of the same material as that tested was also used as a reference in order to follow the variation in the close environment. It was therefore not stretched. The free length of both specimens was equal to 63 mm. Tests were performed at ambient temperature. The loading consisted of load-unload cycles at a displacement rate of $\pm$100 mm/min. Two blocks of 6 cycles were successively applied. The maximum stretch was set to a quarter and a half of the stretch at failure for the first and second block, respectively.

Temperature measurements were performed by using a FLIR infrared camera equipped with a focal plane array of 640 $\times$ 512 pixels and detectors operating in wavelengths between 1.5 and 5.1 μm. The integration time was equal to 1000 μs and the acquisition frequency was the same as for kinematic images, i.e. 25 fps. The thermal resolution or noise equivalent temperature difference is equal to 20 mK for a temperature range between 5 and 40 °C. The infrared camera was switched

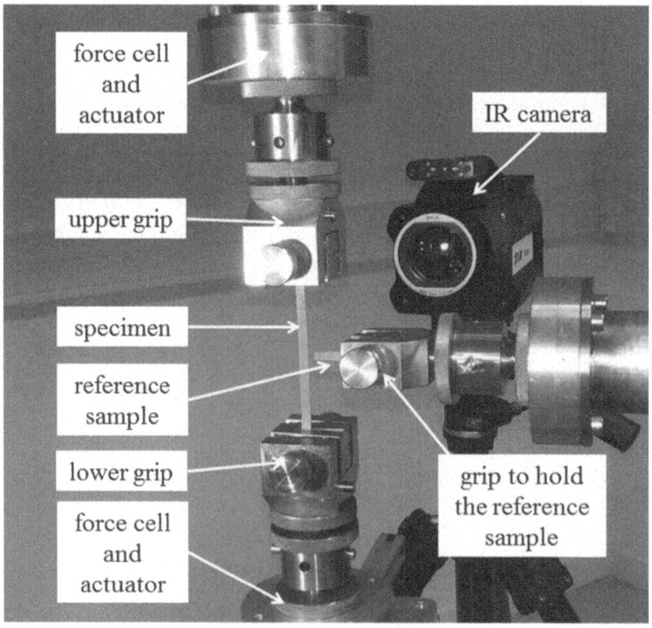

Fig. 12.1 Experimental configuration

on at least 3 h before testing in order to ensure its internal temperature to be stabilized. The calibration of camera detectors was performed with a black body using a Non-Uniformity Correction (NUC) procedure. The variation in mean temperature change of the tested specimen was defined with respect to the reference sample. A spatial averaging over the central zone of the specimen enabled us to improve the thermal measurement resolution.

Tests were conducted with a home-made biaxial testing machine. Four independent RCP4-RA6C-I-56P-4-300-P3-M (IAI) electrical actuators enabled us to stretch symmetrically the samples. They were driven by a PCON CA 56P I PLP 2 0 controller and four PCON-CA (IAI) controller position coupled to a homemade LabView program. The cell load capacity was 1094 N. In the present case, the testing machine was used to stretch the precut sample with two actuators, i.e. with a neutral point in the specimen's middle, which enabled us to avoid complex displacement compensation processing.

12.3 Results

Figure 12.2 shows the mean temperature change as a function of time for both specimens. In both cases, buckling occurred upon unloading in the second block of cycles. Zones corresponding to buckling are highlighted by black rectangles in the graphs. The times corresponding to maximum stretching are indicated by vertical solid lines. Distinct responses are clearly observed. For material L1, temperature was decreasing during the loading phases, and was increasing during the unloading phases. This response is typical of materials featuring an *isentropic coupling*. An opposite behavior is observed for material L2, which thus features an *entropic coupling*. Mechanical dissipation can also be detected from these curves. Indeed for both materials, temperatures was globally increasing over the cycles (*positive* temperature changes), meaning that the total heat produced by the material over the cycles was positive. In other words, the heat produced is higher that the heat absorbed over one mechanical cycle. The next paragraph provides an analysis in terms of heat source variation.

As indicated in the introduction section, the 0D approach was applied to calculate the heat sources in relation to the measured temperatures. Equation (12.1) provides the relation between the heat source s (in W/m^3) and the mean temperature change θ:

$$s = \rho C \left(\frac{d\theta}{dt} + \frac{\theta}{\tau} \right) \tag{12.1}$$

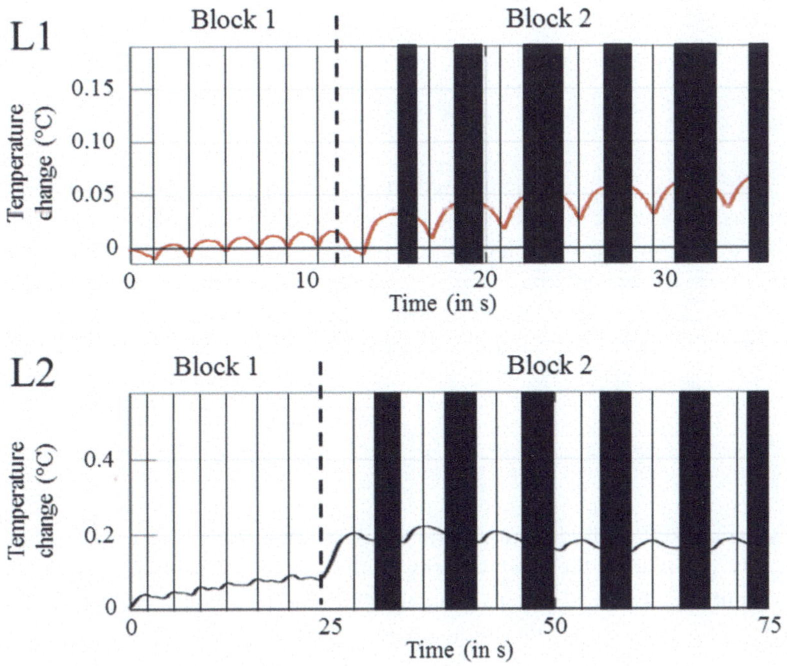

Fig. 12.2 Mean temperature change as a function of time for both specimens. Periods for which specimen buckling occurred are highlighted by black rectangles. The times corresponding to maximum stretching are indicated by vertical solid lines

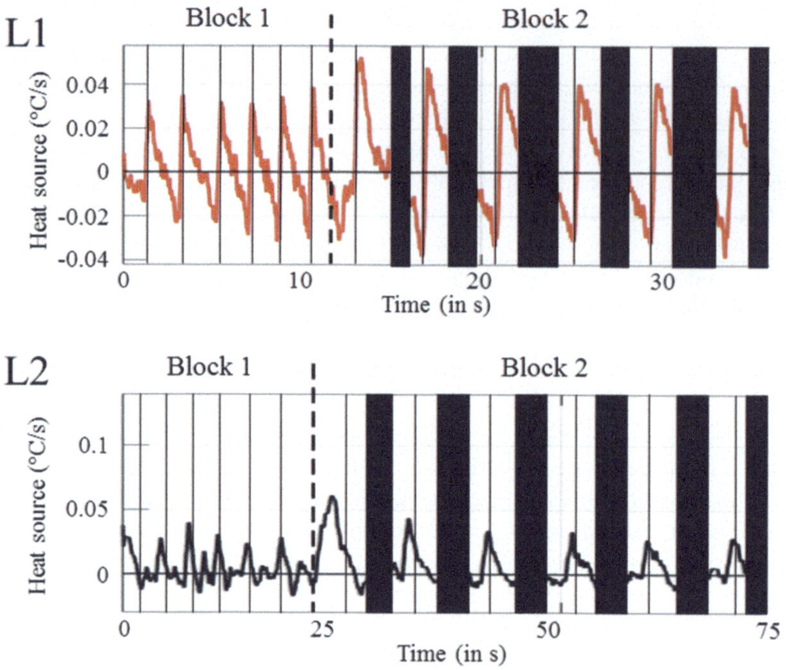

Fig. 12.3 Heat sources as a function of time for both specimens. Periods for which specimen buckling occurred are highlighted by black rectangles. The times corresponding to maximum stretching are indicated by vertical solid lines

where ρ is the material density, C the specific heat and τ a time constant characterizing the global heat exchanges with the outside of the specimen. The value of τ was preliminary measured for each specimen by considering a natural return to ambient temperature. By dividing Eq. (12.1) by ρC, it comes

$$s' = \frac{\mathrm{d}\theta}{\mathrm{d}t} + \frac{\theta}{\tau} \tag{12.2}$$

where

$$s' = \frac{s}{\rho C} \tag{12.3}$$

which can be still named "heat source", but then expressed in °C/s. It corresponds to the temperature rate that would be obtained in adiabatic conditions. Figure 12.3 shows the heat source as a function of time for both specimens. For material L1, heat sources are positive during the loading phases, and negative during the unloading phases, due to an *isentropic coupling*. Opposite signs are observed for material L2, as expected for an *entropic coupling*. By temporal integration over each cycle, it can be shown that the total produced heat is positive. As the heat due to the thermoelastic couplings is null aver a cycle, it means that both materials produced mechanical dissipation.

12.4 Conclusion

Two leathers were tested under uniaxial cyclic loading while temperature changes were measured at the specimens' surface by using IR thermography. By calculating the heat sources at the origin of the temperature changes, the nature of the thermoelastic coupling was identified. Intrinsic dissipation caused by the mechanical irreversibility was also detected. This type of results could be used to propose accurate thermomechanical behavior laws of leathers.

Acknowledgements The authors thank the Campus des Metiers et des Qualifications "Design, Materiaux & Innovation" for supporting this work. The authors also thank the National Center for Scientific Research (MRCT-CNRS and MI-CNRS), Rennes Metropole and Region Bretagne for financially supporting this work. Finally, the authors thank Dr Mathieu Miroir, Dr Eric Robin, Mr Vincent Burgaud and Mr Mickael Lefur for having designed the biaxial tensile machine.

References

1. Mitton, R.G.: Mechanical properties of leather fibres. J. Int. Soc. Leather Trades Chem. **29**, 169–194 (1945)
2. Mitton, R.G.: Tensile properties and their variability in chrome-tanned calfskin. J. Soc. Leather Trades Chem. **32**, 310–323 (1948)
3. Lin, J., Hayhurst, D.R., Howard, I.C., Reedman, D.C.: Modeling of the performance of leather in a uniaxial shoe-last simulator. J. Strain Anal. Eng. Des. **27**, 187–196 (1992)
4. Makho K.: The effect of different parameters on the rupture properties of leather in a tensile test, Master thesis, Faculty of Science, Physics and Electronics, Rhodes University (South Africa) (1998)
5. Manich, A.M., de Castellar, M.D., Gonzalez, B., Ussman, M.H., Marsal, A.: Influence of leather stretching to gain area yield on its stress-relaxation behavior. J. Appl. Polym. Sci. **102**, 6000–6008 (2006)
6. Bison, P.G., Grinzato, E., Marinetti, S.: Leather characterisation by IR thermography. In: Peacock, G.R., Burleigh, D.D., Miles, J.J. (eds.) Proceedings of the Society of the Photo-Optical Instrumentation Engineers (SPIE), vol. 5782, pp. 359–370. SPIE-Int Soc Optical Engineering, Bellingham, WA, USA (2005)
7. Luong, M.P.: Evaluation of the limit of acceptable damage for leather products using infrared thermography. In: Baaklini, G.Y., Lebowitz, C.A., Boltz, E.S. (eds.) Nondestructive Evaluation of Aging Materials and Composites III, Proc. SPIE, vol. 3585, pp. 84–95. SPIE-Int Soc Optical Engineering, Bellingham, WA, USA (1999)
8. Luong, M.P.: Evaluating a limit of acceptable damage for leather products. In: Ellyin, F., Provan, J.W. (eds.) Progress in Mechanical Behavior of Materials (ICM8), Univ Victoria, Dept Mech Engn, Victoria, Canada, vol. 2, Materials properties, pp. 523–528 (1999)
9. Luong, M.P.: Infrared thermography of damage evaluation in leather products. In: Melker, A.I. (ed.) Proceedings of the Society of the Photo-Optical Instrumentation Engineers (SPIE), vol. 3687, pp. 191–200. SPIE-Int Soc Optical Engineering, Bellingham, WA, USA (1999)
10. Chrysochoos, A., Dupré, J.C.: Experimental analysis of thermomechanical coupling by infra-red thermography. In: Boehler, J.P., Khan, J.P. (eds.) Anisotropy and localization of plastic deformation, pp. 540–543. Springer, Dordrecht (1991)
11. Meyer, K.H., Ferri, C.: Sur l'élasticité du caoutchouc. Helv. Chim. Acta. **18**, 570–589 (1935)
12. Treloar, L.R.G.: The elasticity and related properties of rubbers. Rep. Prog. Phys. **36**, 755–826 (1973)
13. Chadwick, P.: Thermo-mechanics of rubberlike materials. Philos. Trans. Royal Soc. A Math. Phys. Eng. Sci. **276**, 371–403 (1974)
14. Chrysochoos, A.: Analyse du comportement des matériaux par thermographie infrarouge. In: Berthaud, Y. (ed.) Photomécanique 95, Cachan, France, 14–16 March, 1995, pp. 203–211. Eyrolles, Paris (1995)
15. Chrysochoos, A., Louche, H.: An infrared image processing to analyse the calorific effects accompanying strain localization. Int. J. Eng. Sci. **38**, 1759–1788 (2000)
16. Samaca Martinez, J.R., Le Cam, J.B., Balandraud, X., Toussaint, E., Caillard, J.: Mechanisms of deformation in crystallizable natural rubber. Part 1: thermal characterization. Polymer. **54**, 2727–2736 (2013)
17. Samaca Martinez, J.R., Le Cam, J.B., Balandraud, X., Toussaint, E., Caillard, J.: Filler effects on the thermomechanical response of stretched rubbers. Polym. Test. **32**, 835–841 (2013)
18. Samaca Martinez, J.R., Le Cam, J.B., Balandraud, X., Toussaint, E., Caillard, J.: New elements concerning the Mullins effect: a thermomechanical analysis. Eur. Polym. J. **55**, 98–107 (2014)
19. Balandraud, X., Le Cam, J.B.: Some specific features and consequences of the thermal response of rubber under cyclic mechanical loading. Arch. Appl. Mech. **84**, 773–788 (2014)
20. Le Cam, J.B., Samaca Martinez, J.R., Balandraud, X., Toussaint, E., Caillard, J.: Thermomechanical analysis of the singular behavior of rubber: entropic elasticity, reinforcement by fillers, strain-induced crystallization and the Mullins effect. Exp. Mech. **55**, 771–782 (2015)
21. Le Cam, J.B.: Energy storage due to strain-induced crystallization in natural rubber: the physical origin of the mechanical hysteresis. Polymer. **127**, 166–173 (2017)
22. Le Cam, J.B.: Strain-induced crystallization in rubber: a new measurement technique. Strain. **54**, e12256 (2018)
23. Chrysochoos, A., Pham, H., Maisonneuve, O.: Energy balance of thermoelastic martensite transformation under stress. Nucl. Eng. Des. **162**, 1–12 (1996)
24. Balandraud, X., Ernst, E., Soos, E.: Rheological phenomena in shape memory alloys. Comptes-Rendus de l'Académie des Sciences - Série II Fascicule B Mécanique Physique Astronomie. **327**, 33–39 (1999)
25. Balandraud, X., Ernst, E., Soos, E.: Relaxation and creep phenomena in shape memory alloys. Part II: stress relaxation and strain creep during phase transformation. Z. Angew. Math. Phys. **51**, 419–448 (2000)
26. Balandraud, X., Ernst, E., Soos, E.: Relaxation and creep phenomena in shape memory alloys. Part I: hysteresis loop and pseudoelastic behavior. Z. Angew. Math. Phys. **51**, 171–203 (2000)
27. Bubulinca, C., Balandraud, X., Grédiac, M., Stanciu, S., Abrudeanu, M.: Characterization of the mechanical dissipation in shape-memory alloys during stress-induced phase transformation. J. Mater. Sci. **49**, 701–709 (2014)
28. Delobelle, V., Favier, D., Louche, H., Connesson, N.: Determination of local thermophysical properties and heat of transition from thermal fields measurement during drop calorimetric experiment. Exp. Mech. **55**, 711–723 (2015)
29. Delpueyo, D., Balandraud, X., Grédiac, M., Stanciu, S., Cimpoesu, N.: A specific device for enhanced measurement of mechanical dissipation in specimens subjected to long-term tensile tests in fatigue. Strain. **54**, e12252 (2017)
30. Benaarbia, A., Chrysochoos, A., Robert, G.: Kinetics of stored and dissipated energies associated with cyclic loadings of dry polyamide 6.6 specimens. Polym. Test. **34**, 155–167 (2014)

31. Benaarbia, A., Chrysochoos, A., Robert, G.: Influence of relative humidity and loading frequency on the PA6.6 cyclic thermomechanical behavior: Part I. Mechanical and thermal aspects. Polym. Test. **40**, 290–298 (2014)
32. Benaarbia, A., Chrysochoos, A., Robert, G.: Thermomechanical behavior of PA6.6 composites subjected to low cycle fatigue. Composites Part B-Eng. **76**, 52–64 (2015)
33. Benaarbia, A., Chrysochoos, A., Robert, G.: Influence of relative humidity and loading frequency on the PA6.6 thermomechanical cyclic behavior: part II. Energy aspects. Polym. Test. **41**, 92–98 (2015)
34. Benaarbia, A., Chrysochoos, A., Robert, G.: Fiber orientation effects on heat source distribution in reinforced polyamide 6.6 subjected to low cycle fatigue. J. Eng. Math. **90**, 13–36 (2015)
35. P. Jongchansitto, C. Douellou, I. Preechawuttipong, X. Balandraud, Comparison between 0D and 1D approaches for mechanical dissipation measurement during fatigue tests, strain, (submitted 2018)
36. Giancane, S., Chrysochoos, A., Dattoma, V., Wattrisse, B.: Deformation and dissipated energies for high cycle fatigue of 2024-T3 aluminium alloy. Theorerical Appl. Fracture Mech. **52**, 117–121 (2009)
37. Boulanger, T., Chrysochoos, A., Mabru, C., Galtier, A.: Calorimetric analysis of dissipative and thermoelastic effects associated with the fatigue behavior of steels. Int. J. Fatigue. **26**, 221–229 (2004)
38. Wang, X.G., Crupi, V., Jiang, C., Feng, E.S., Guglielmino, E., Wang, C.S.: Energy-based approach for fatigue life prediction of pure copper. Int. J. Fatigue. **104**, 243–250 (2017)

Chapter 13
Experimentally Enhanced Computations: Calibration Methodology for an Anisotropic Metal, Part II – Novel Approach/Validation

Sharlotte Kramer, Amanda Jones, Brian Lester, and Edmundo Corona

Abstract Here, we present our current progress towards Part II of the Experimentally Enhanced Computations project: a novel calibration approach. While the first part discussed the traditional calibration approach, the availability of advanced diagnostics combined with the development of a new finite element updating inverse method that utilizes full field displacement data and the virtual fields method enables a novel calibration approach.

Keywords Model calibration · Computational modeling · Anisotropy · Digital image correlation · Aluminum

Two specimen geometries are explored: an offset notched tensile bar in tension and a modified Brazilian disk tested in compression, both being machined at several orientations from the same aluminum extrusion used in Part I. Figure 13.1 shows images of the experimental setups used for both specimen geometries. Digital image correlation (DIC) was used to measure the displacement fields on both the front and back sides of the specimens.

For the first geometry, double notched tension, specimens were pulled in tension at a rate of 5 μm/s until failure. The orientation with respect to the extrusion axis factored into the crack path of the notched tensile specimens, and was evident in the strain fields. One representative DIC result for this specimen geometry is given in Fig. 13.2a. Similarly, for the second geometry, the Brazilian disk, specimens with 20 mm nominal diameter were compressed to a maximum displacement of 1 mm (Fig. 13.2b). Brazilian specimens machined in one of the orientations consistently fractured during loading, while those machined in the other orientations were compressed to the full 1\,mm of prescribed displacement without fracture.

The results from the tests serve two purposes in our work: the principal one is to provide the data necessary to implement the novel calibration approach and the other is to compare the test data to results of simulations that use the traditional calibration approach. As an example, preliminary results for the transverse strain (perpendicular to the axis of pull) from double notch tension specimens cut from two orientations (labeled A and B) are shown in Fig. 13.3 using the calibration parameters from a Barlat model. This model captures a difference between the two orientations, which is mirrored in the DIC results.

The summary of these comparisons in terms of both the modeling approaches and material orientations, will be presented first, followed by a discussion of the methodology of the novel approach and the validation procedure that we envision.

S. Kramer · A. Jones (✉) · B. Lester · E. Corona
Sandia National Laboratories, Albuquerque, NM, USA
e-mail: ajones1@sandia.gov

A. Baldi et al. (eds.), *Residual Stress, Thermomechanics & Infrared Imaging, Hybrid Techniques and Inverse Problems, Volume 7*, Conference Proceedings of the Society for Experimental Mechanics Series, https://doi.org/10.1007/978-3-319-95074-7_13

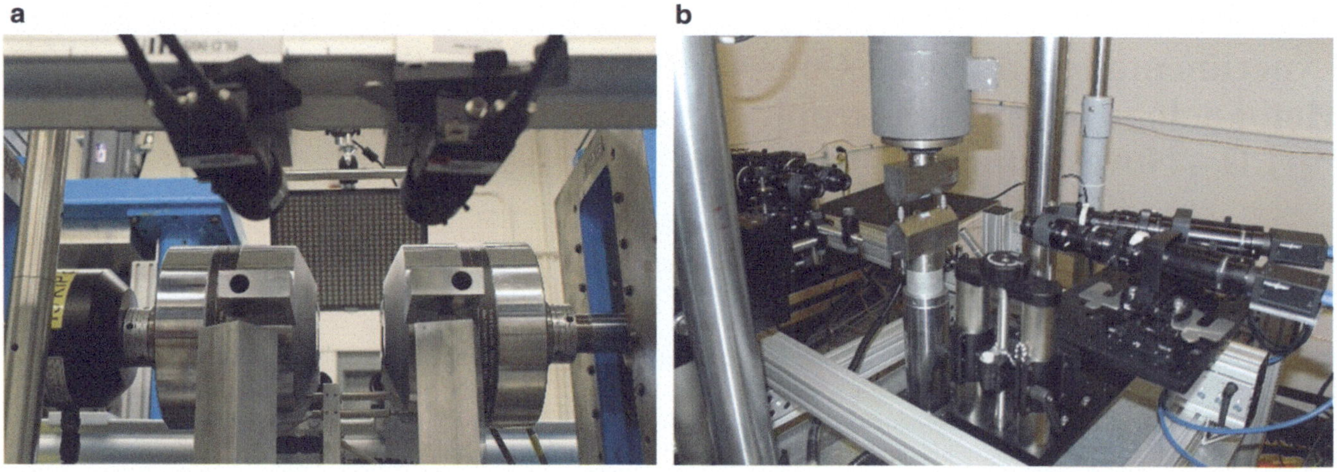

Fig. 13.1 (**a**) Experimental setup used for the double notch tension tests and (**b**) experimental setup used for the modified Brazilian disk tests. Note four cameras are used in both setups to capture deformation on both the front and back sides of the specimens

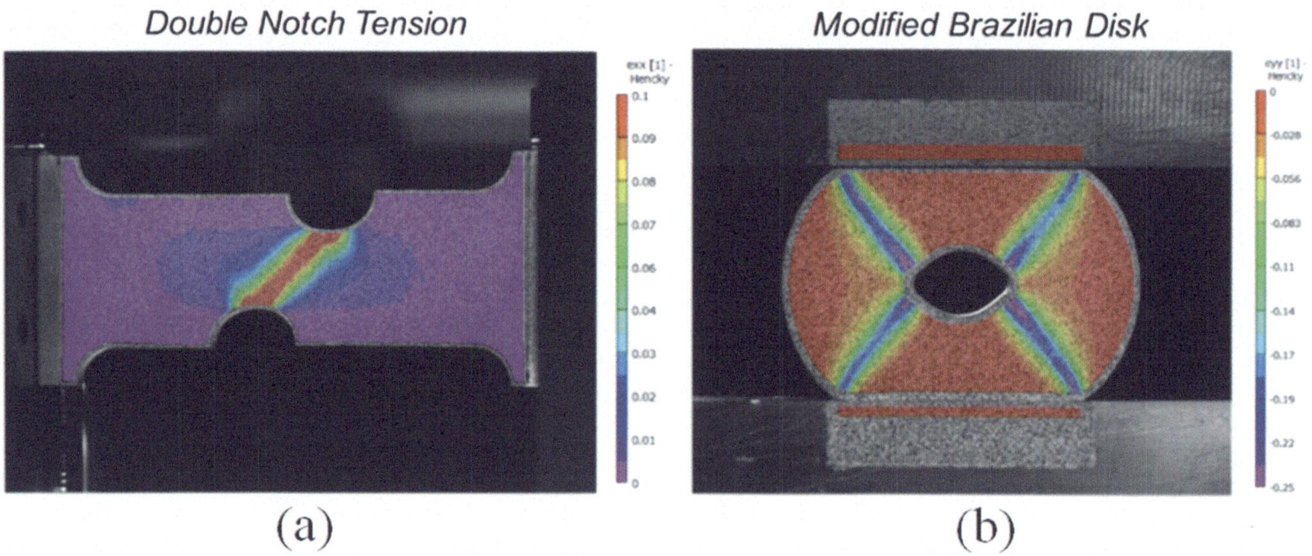

Fig. 13.2 Representative DIC results from the two novel geometries right before the completion of the test for (**a**) double notch tension and (**b**) modified Brazilian disk. Strain fields are given for the direction parallel to deformation (tension and compression respectively)

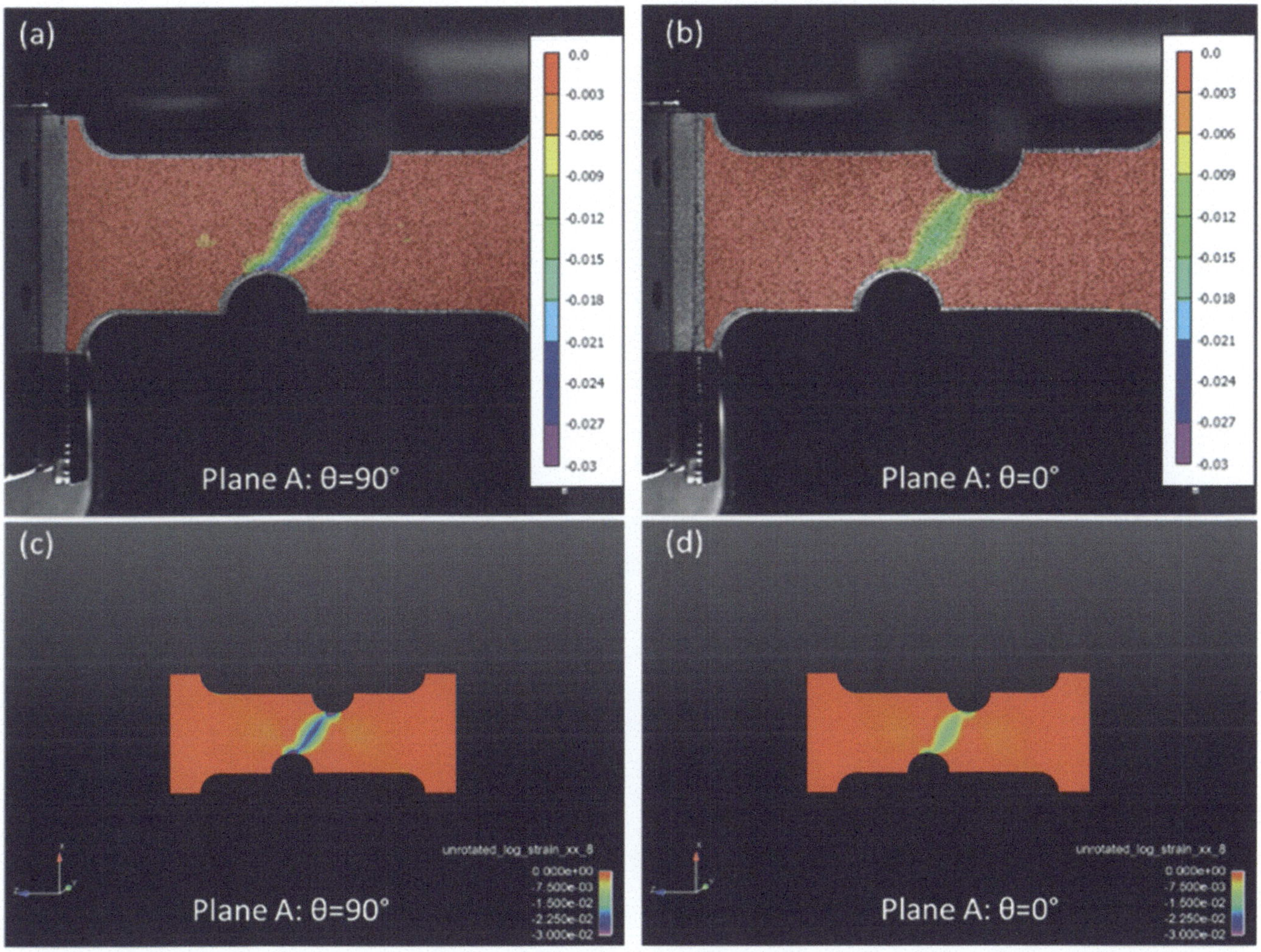

Fig. 13.3 Comparison of transverse logarithmic (Hencky) strains at the total travel of 0.64 mm from DIC from a specimen aligned in (**a**) Plane A, $\theta = 90°$ and (**b**) Plane A, $\theta = 0°$ to computational predictions using a Barlat anisotropy yield function for (**c**) Plane A, $\theta = 90°$ and (**d**) Plane A, $\theta = 0°$

Acknowledgements Sandia National Laboratories is a multimission laboratory managed and operated by National Technology and Engineering Solutions of Sandia, LLC., a wholly owned subsidiary of Honeywell International, Inc., for the U.S. Department of Energy's National Nuclear Security Administration under contract DE-NA-0002525.

Chapter 14
Multi-instrumentation of Very High Temperature Tests

T. Archer, P. Beauchêne, M. Berny, and F. Hild

Abstract For the use of thermal and environmental barrier coating (T/EBC) with ceramic matrix composite it is crucial to understand the behavior under extreme environments representative of the hot sections of turbine engines. An experimental setup to simulate such thermal loading has been developed with various instrumentation enabling for kinematic and temperature field measurements.

Keywords Environmental barrier coating · Ceramic matrix composite · Infrared thermography · Digital image correlation · Heat haze effect

14.1 Introduction

Ceramic matrix composites (CMCs) are currently considered as an attractive option for use in the hot sections of next generation turbine engines. Their high temperature mechanical and physical properties and their low density could provide increased performance under such extreme environments. Among the CMCs developed today, silicon carbide (SiC) fiber-reinforced melt-infiltrated (MI) SiC matrices have shown good stability up to 1300 °C [1], especially against high temperature oxidation in dry air. However, for applications in turbine engines like blades or shroud, the presence of water vapor leads to surface recession of the protective silica layer formed at high temperatures [2]. It is therefore essential to protect the CMC against operating temperatures but also corrosive environments, which calls for the development of new thermal and environmental barrier coatings (T/EBCs).

In order to successfully develop T/EBC-CMC systems, the understanding of the behavior of T/EBC under representative stresses, temperatures and corrosive atmospheres is essential. Given the environment, these components will face high thermal gradients due to high heat fluxes and forced air cooling resulting in complex thermomechanical stress states.

The differences between isothermal and thermal gradient conditions [3] make standard tests used for characterization ill-adapted to assess damage initiation and growth. For that reason a test rig has been developed to examine T/EBC-CMC systems (Fig. 14.1).

T. Archer (✉)
Office National d'Etudes et de Recherches Aérospatiales (ONERA), Palaiseau, France

Laboratoire de Mécanique et Technologie (LMT), ENS Paris-Saclay, Cachan, France

SAFRAN, Safran Ceramics, Le Haillan, France
e-mail: thibaut.archer@onera.fr

P. Beauchêne
Office National d'Etudes et de Recherches Aérospatiales (ONERA), Palaiseau, France

M. Berny
Laboratoire de Mécanique et Technologie (LMT), ENS Paris-Saclay, Cachan, France

SAFRAN, Safran Ceramics, Le Haillan, France

F. Hild
Laboratoire de Mécanique et Technologie (LMT), ENS Paris-Saclay, Cachan, France

A. Baldi et al. (eds.), *Residual Stress, Thermomechanics & Infrared Imaging, Hybrid Techniques and Inverse Problems, Volume 7*,
Conference Proceedings of the Society for Experimental Mechanics Series, https://doi.org/10.1007/978-3-319-95074-7_14

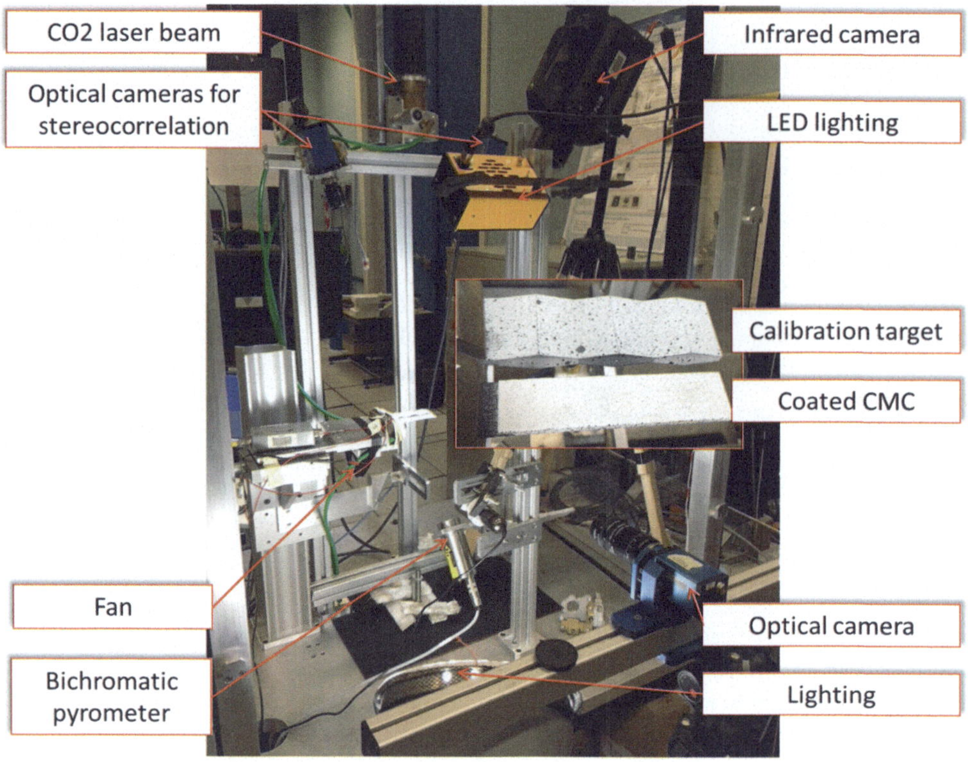

Fig. 14.1 High heat flux laser test using visible light and infrared cameras

14.2 Experimental Setup and Tested Materials

A high power (2 kW) high heat flux CO2 laser heats locally the sample, thereby generating multi-axial (in-plane and through-thickness) thermal gradients. Under such conditions, the material expands and deflects. To measure the thermomechanical fields of the sample, the top surface is monitored using a middle-wave infrared (IR) camera (FLIR X6580sc) to measure temperature fields, a pair of optical cameras (AVT Pike-421 with 2.8/50 lenses) and an LED lighting (7700 Lumen) to measure 3D surface displacement fields. The back side is monitored using bichromatic (IGAR6, Lumasens) and monochromatic (MI3, Raytek) pyrometers. An additional optical camera (AVT Pike-421 with 4/80 lens and 41 mm extension ring) is positioned to measure the edge motion of the sample.

The tested sample consists of a woven CMC manufactured with Hi-Nicalon S SiC fiber and SiC matrix. The substrate is then coated with a thin layer of bond coat, and finally with an environmental barrier coating.

14.3 Full-Field Measurements

Given the complex manufacturing process of T/EBC, the knowledge of thermomechanical properties is often complicated and expensive when the material undergoes heat-treatments that produce physical and/or structural changes.

During high temperature tests, an accurate measurement of temperature and displacement fields will allow in-situ estimations of key thermomechanical parameters such as the thermal expansion coefficient and Young's modulus, which play a critical role in the thermomechanical behavior of T/EBC-CMC systems. Therefore, for the monitoring of such tests, a specific instrumentation needs to be implemented.

IR thermography is performed with a middle wave infrared camera [3 μm; 5 μm] with a band-pass filter at 4 μm. It is sensitive to a radiant flux coming from the thermal scene, which is converted into digital levels (DL) and then with the emissivity into temperature. The advantage of a monochromatic camera is that the radiant flux can be expressed with Planck's emission law. Given the large range of temperatures during laser heating and the change of emissivity ϵ as a function of the

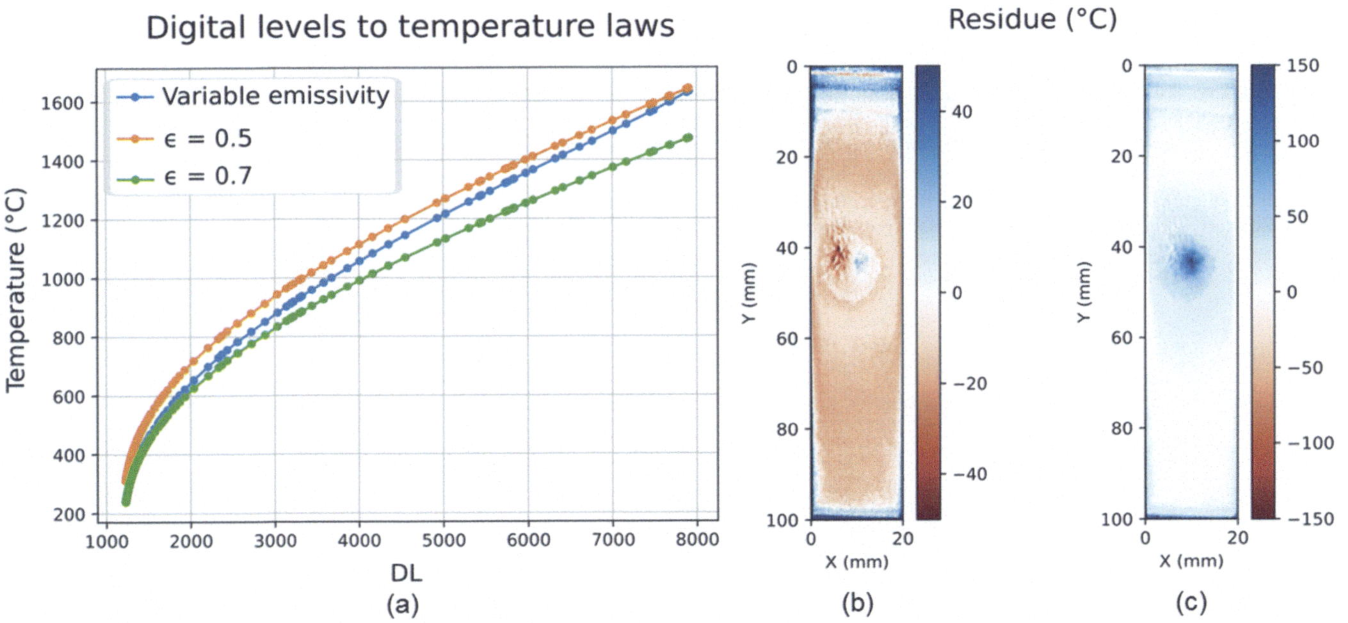

Fig. 14.2 (**a**) Temperature calibration laws, residue between field rebuilt with variable emissivity and constant emissivity (**b**) $\epsilon = 0.5$ and (**c**) $\epsilon = 0.7$

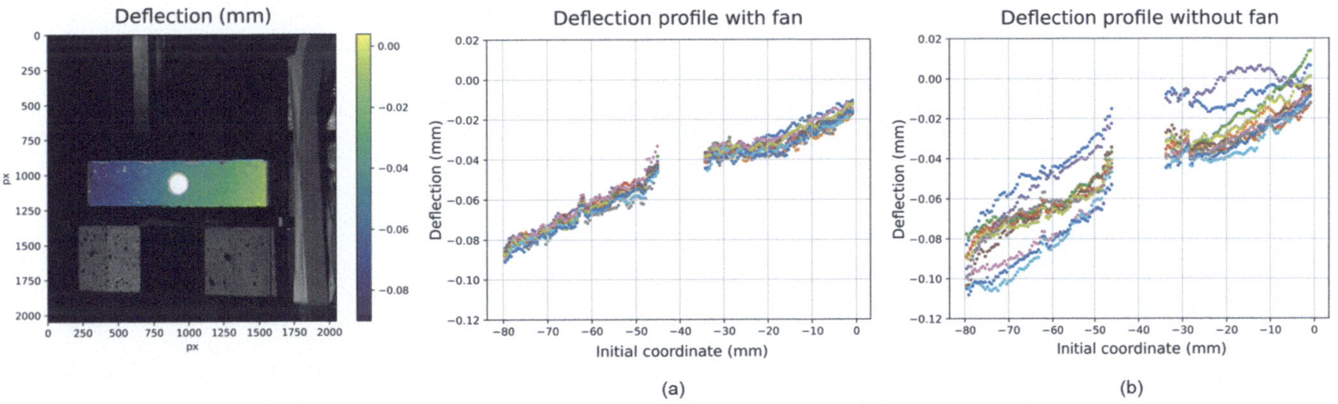

Fig. 14.3 Out of plane measurement (**a**) with fan (**b**) with no fan

temperature, the rebuilt of the temperature field with a constant emissivity induces errors (Fig. 14.2). Using a low emissivity will overestimate low temperatures, when using a high emissivity will largely underestimate high temperatures.

Under such thermal loadings, the system expands and deflects. One of the biggest challenges is to measure very low displacement levels (a few tens of micrometers). For the use of stereocorrelation, a high temperature speckle pattern was created using SiC powder. A fan was used to reduce heat haze effects and lower the uncertainties. Visible light images were post-processed using the commercial software VIC 3D (Correlated Solutions). Results of 10 pictures taken every 10 s from a stabilized temperature step are compared with and with no fan (Fig. 14.3). The white zone at the center of the image corresponds to the laser beam (i.e., the highest temperatures) where the pixels of the optical camera are saturated. In the present case, the use of a fan is critical in reducing the biases induce by the heat haze.

This work was supported under PRC MECACOMP, French research project co-funded by DGAC and SAFRAN Group, piloted by SAFRAN Group and involving SAFRAN Group, ONERA and CNRS.

References

1. DiCarlo, J.A., Yun, H.-M., Morscher, G.N., Bhatt, R.T.: SiC/SiC composites for 1200°C and above. In: Handbook of Ceramic Composites, pp. 77–98 (2005)
2. More, K.L., Tortorelli, P.F., Walker, L.R., Miriyala, N., Price, J.R., Roode, M.: High temperature stability of SiC-based composites in high-water-vapor-pressure environments. J. Am. Ceram. Soc. **86**(8), 1272–1281 (2003)
3. Tan, Y., Longtin, J.P., Sampath, S., Zhu, D.: Temperature-gradient effects in thermal barrier coatings: an investigation through modeling, high heat flux test, and embedded sensor. J. Am. Ceram. Soc. **93**(10), 3418–3426 (2010)

Chapter 15
Detection of Damage During Quasi-Static Loading of a Single Stringer Panel Using Passive Thermography and Acoustic Emission

Joseph N. Zalameda, William P. Winfree, and Michael R. Horne

Abstract The purpose of this paper is to describe the methodology of measuring damage onset and growth in a composite structure during quasi-static loading using passive thermography and acoustic emission. The early detection and measurement of damage progression is important to understand failure modes. A single stringer panel was subjected to quasi-static loading to induce deformation which resulted in the formation of damage between the stiffener flange and skin. The loading was stopped when damage growth was detected. Passive thermography and acoustic emission were used to detect damage in real-time as a function of the applied load. Of particular interest are the small transient thermographic signals resulting from damage formation which can be challenging to detect, as compared to the persistent passive thermography indications of cyclic fatigue loading. We describe a custom developed thermal inspection system for detection of composite damage during quasi-static loading. The thermal results are compared to a two-dimensional multi-layered thermal simulation based on the quadrupole method. Acoustic emission is used to further characterize the damage by comparing the acoustic emission events with the thermal imagery. Results are compared to ultrasonic measurements to document the damage through-the-thickness.

Keywords Composite inspection · Thermography · Acoustic emission · Fiber break · Matrix crack · Delamination · Quasi-static loading

15.1 Introduction

Real-time nondestructive evaluation (NDE) is necessary for composites load testing to track early onset and growth of damage. The NDE allows for controlling the growth of the damage as a function of the applied load. In this work, a quasi-static bending load (using 7 contact points) is applied to a single stringer stiffened composite panel. When damage is detected with the real-time NDE, the loading is stopped and another inspection technique such as ultrasound or X-ray CT is used to provide a detailed assessment of the panel damage as a function of depth. Documentation of damage onset, growth, and panel failure will provide valuable information for development and validation of progressive damage analysis (PDA) models [1–3]. The ultimate goal is to use the validated PDA models to decrease the time required to test and certify composite materials/designs for aircraft structures and therefore save development costs. The purpose of this paper is to describe the methodology of measuring damage onset and growth in a composite structure during quasi-static loading using passive thermography and acoustic emission.

There have been numerous studies combining thermography with acoustic emission for structures testing with most applications for fatigue loading [4–9]. The challenge for passive thermography is to detect the small transient thermal signals generated by a quasi-static load as compared to the repetitive thermal signals resulting from cyclic loading. Similar to previous studies, we temporally compare the acoustic emission event with the thermal imagery. This allows confirmation that a detected acoustic emission event, confirmed by the thermal imagery providing location, results in damage onset or growth that would be of interest to characterize using ultrasound or X-ray CT. The thermal anomalies are very small and barely within the detection limits of the infrared camera and therefore a custom data acquisition system was developed to

J. N. Zalameda (✉) · W. P. Winfree
NASA Langley Research Center, MS231, Hampton, VA, USA
e-mail: joseph.n.zalameda@nasa.gov

M. R. Horne
National Institute of Aerospace, NASA Langley Research Center, MS231, Hampton, VA, USA

A. Baldi et al. (eds.), *Residual Stress, Thermomechanics & Infrared Imaging, Hybrid Techniques and Inverse Problems, Volume 7*, Conference Proceedings of the Society for Experimental Mechanics Series, https://doi.org/10.1007/978-3-319-95074-7_15

detect the small thermal signatures. Multiple IR cameras and multiple acoustic emission sensors could be employed for large scale testing of structures, however for this effort given the panel size and setup a single thermal camera and six acoustic emission sensors were utilized.

15.2 Sample

The stiffened composite panel skin is 12 plies with a thickness of 0.22 cm, the stiffener flange is 12 plies thick with a thickness of 0.24 cm. The stiffener is a woven composite. The stiffened composite panel flat side with acoustic emission locations, stringer side, the painted specimen, and cross-sectional view are shown in Fig. 15.1a–d respectively. The panel was painted for digital image correlation measurements to record panel deformation. Quasi-static loads were applied using seven application points, two on top (located in middle just outside of the flange) and five on the bottom (located at each corner and center). The load was applied from the bottom while the top was held stationary at the two points. This configuration allows for panel deformation that results in damage formation between the stiffener flange and skin. The applied quasi-static loads were up to 1000 pounds. Passive thermography was used to track damage on the stringer side and the array of acoustic emission sensors were mounted on the skin side. The test setup is shown in Fig. 15.2a along with infrared camera inspection view shown in Fig. 15.2b. Examples of the applied load for a typical test and panel setup while under load showing deflection are shown in Fig. 15.3a, b respectively.

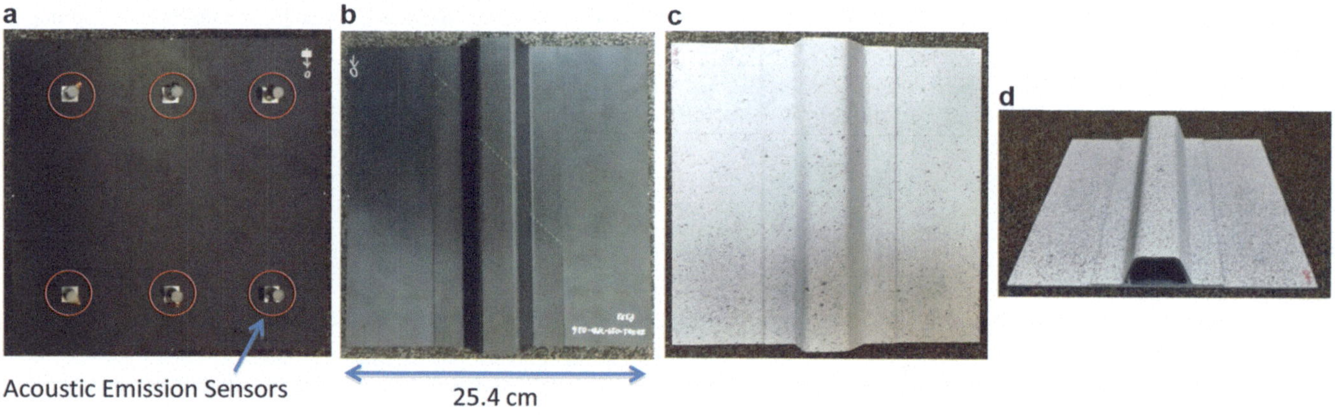

Fig. 15.1 Single stringer composite panel with acoustic emission sensor locations. (**a**) Flat side, (**b**) Stringer side, (**c**) Stringer side (painted), (**d**) Cross sectional view

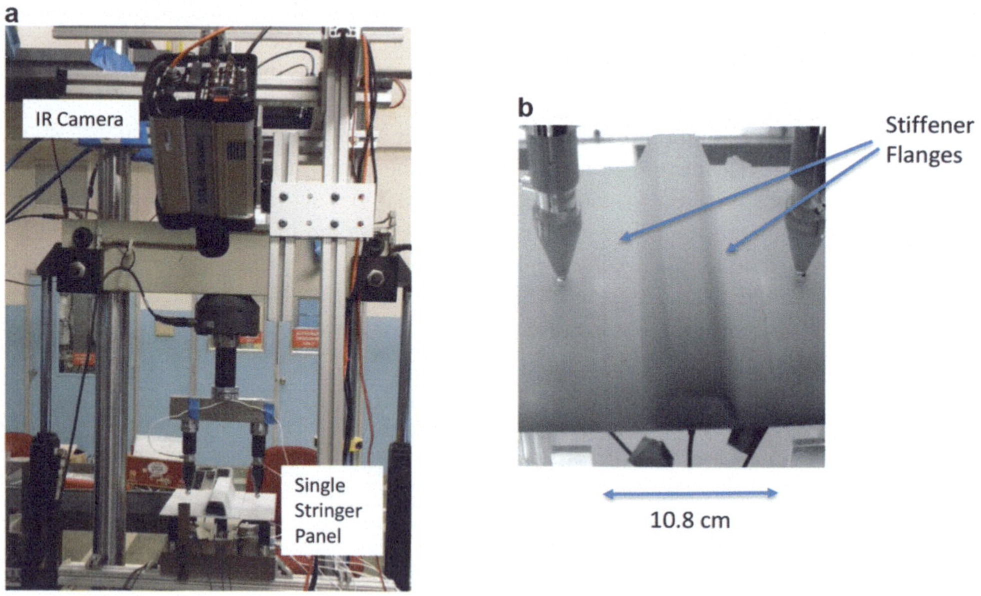

Fig. 15.2 Fatigue loading setup for single stringer panel with infrared camera view. (**a**) Setup, (**b**) Infrared camera view

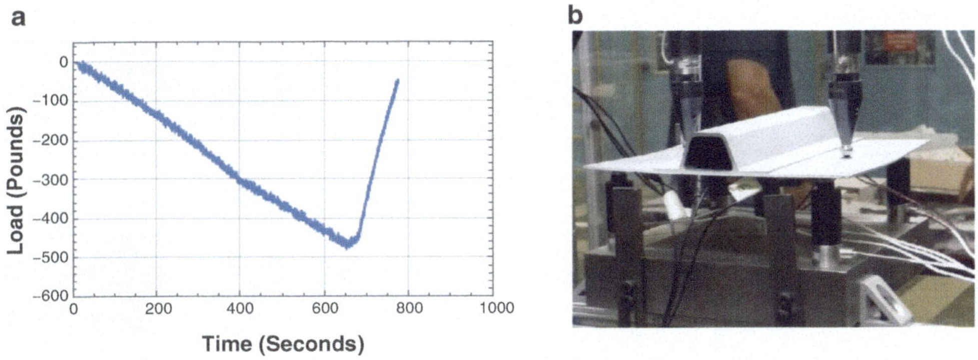

Fig. 15.3 Applied load and panel deformation. (**a**) Quasi-static loading, (**b**) panel deformation

15.3 Inspection Systems

The IR camera shown in Fig. 15.2 operates in the mid-wave 3–5 micrometer band. The IR camera is configured with a 25 mm germanium lens. The focal plane array size for the camera is 640 × 512. The camera frame rate was from 80 to 180 Hz. The applied load was acquired simultaneously with the thermal imagery at the same sampling rate of 80 Hz. Both the unprocessed thermal images and applied load were written to a solid state hard drive while displaying the processed imagery. Real-time processing of the displayed thermal images was required to detect the small thermal indications of damage. This involved applying a boxcar averaging and a delayed image difference technique in real-time [9, 10]. The setup required a Plexiglas® shield (not shown in Fig. 15.2a) to filter out spurious IR background sources.

Six acoustic emission sensors were bonded to the composite panel on the skin side (shown in Fig. 15.1a). Each sensor was connected to a data acquisition computer via Digital Wave ™ PA0 preamp/line drivers and a Digital Wave ™ FM1 signal conditioning channel amplifier for capture of acoustic data at 2 MHz sample rate and 1024 points per event. The acoustic emission computer also captured the load for synchronization to the thermal images. Acoustic emission systems collect structure-borne sound, typically not audible, in the ultrasonic frequency band (approx. 50–500 kHz), generated by dynamic displacements such as damage initiation and growth at picometer scales [11]. Pre-cursors to failure are monitored in real-time. The total signal energy, *SE*, of an acoustic signal is given as:

$$SE = \sum_{i=1}^{N} V_i^2 \Delta t \tag{15.1}$$

where V_i is the signal voltage, i is the time reference point, N is the number of time points in the signal, Δt is the sampling time per point. Trends of the energy over time, or load, are used to identify and track damage development. Before each load test, pencil lead breaks are performed at various locations on the panel to calibrate and test the acoustic emission sensors.

Ultrasonic inspections were implemented at intervals determined by the indication of damage growth from thermography and acoustic emission. The ultrasonic inspection technique utilizes a broadband 10 MHz transducer (1.27 cm diameter with 5.08 cm focus) contained in a captive water column with a flexible membrane tip design [12]. Water is sprayed onto the surface for ultrasonic coupling. The probe is scanned over the entire panel on the flat side. A picture of the setup is shown in Fig. 15.4.

15.4 Measurement Results

Digital image processing was required to both enhance the detection of thermal events during load and to facilitate comparison of the thermal inspection imagery to the ultrasonic data. To enhance detection in real-time, image averaging and a delayed subtraction algorithm was implemented. Typical parameters of 20–100 frames were averaged and a delay subtraction of 20 averaged frames were typically used. The delayed subtraction removed fixed background infrared radiation while increasing sensitivity to changes. To further remove noise, a de-noising algorithm using singular value decomposition was implemented [13, 14]. Lastly, an image perspective transformation was used to correct for the infrared camera look angle since the optical line of sight was not normal [14, 15]. Detected damage progression during load runs 1–4 are shown

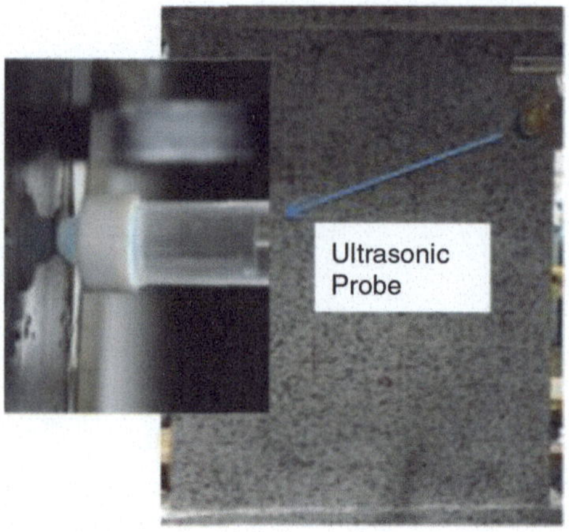

Fig. 15.4 Close up of ultrasonic probe

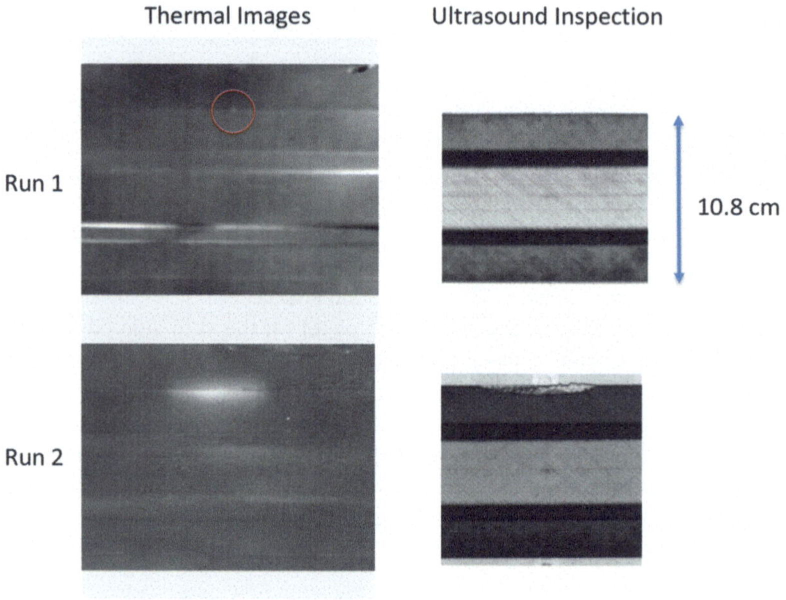

Fig. 15.5 Processed thermal inspection images for quasi-static load runs 1 and 2 with corresponding ultrasound inspection images

in Figs. 15.5, 15.6, and 15.7. The infrared image is rotated right so the top stiffener flange is the left flange of Fig. 15.2b. The thermal damage was detected over camera noise and background reflections using three criteria: observing a bright area that appears suddenly, stays in one location, and then slowly decays in intensity over time. When damage is detected the loading is stopped, the sample is removed, and an ultrasound inspection is performed to document the damage through-the-thickness. The respective ultrasound inspections are also shown in Figs. 15.5, 15.6, and 15.7. In Fig. 15.5, for run 1, a small thermal indication (circled in red) was detected and the resulting ultrasonic inspection revealed no damage. This was typical for several of the samples tested. A small thermal indication did not always result in detectable damage using ultrasound. The small indication in Fig. 15.5 could be a small matrix crack or fiber break located at the stiffener flange edge perhaps beyond the detection limits of ultrasound. Also, any damage at this edge interface would be difficult to detect with ultrasound especially at the corner. X-ray CT inspection could be used to help understand these initial small thermal indications. The run 2 thermal image in Fig. 15.5 clearly shows a delamination at the stiffener flange edge. The thermal energy release due to the disbond is sufficient to generate a temperature increase to provide the general shape of the damage. The ultrasound inspection image is in agreement with the thermal indication. For load run 3 it is interesting to note the delamination growth, as indicated by the ultrasound inspection image, did not produce a sufficient temperature increase to resolve the shape and size of the damage. This is due to the likely-hood the delamination grew in increments from one side to another. This is plausible since the small thermal indications (circled) were detected at opposite sides of the delamination at different times

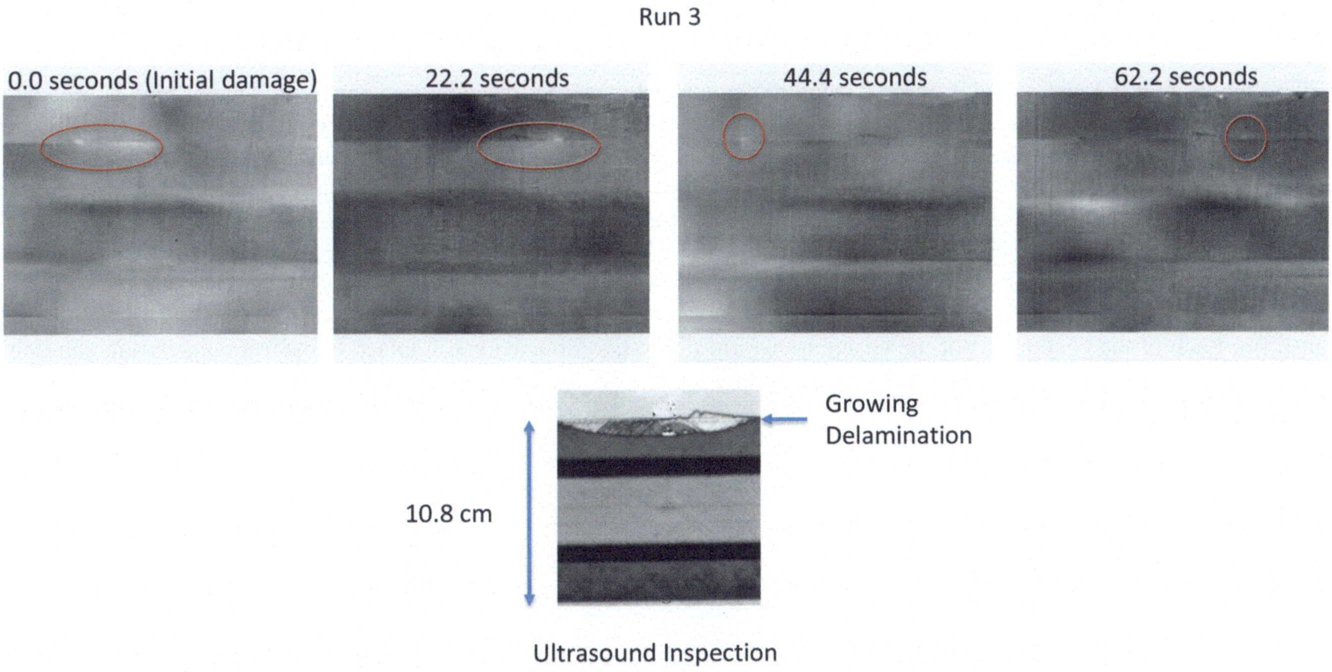

Fig. 15.6 Processed thermal inspection images for quasi-static load run 3 with corresponding ultrasound inspection image

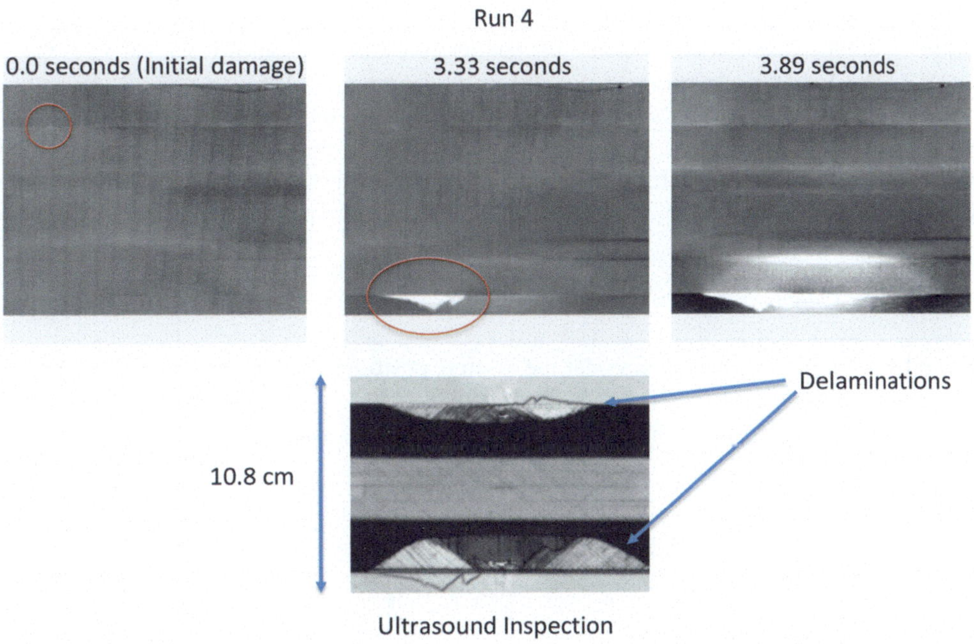

Fig. 15.7 Processed thermal inspection images for quasi-static load run 4 with corresponding ultrasound inspection image

as shown in Fig. 15.6. Shown in Fig. 15.7 is the final load run 4 where a large delamination was detected at the bottom as shown in the ultrasound inspection image. From the thermal inspection imagery, the top delamination appeared to grow only slightly on the left side. At 3.33 s, there appears to be a single layer failure (circled) at the skin leading into the stiffener flange and the faint outline of the delamination. At 3.88 s, there is a large thermal indication that appears more clearly showing the overall shape of the delamination with good agreement as shown by the ultrasound inspection.

15.5 Thermal Modeling

A two-dimensional thermal model is studied using the quadrupole method presented previously in a paper by Winfree and verified by comparison to finite element modeling [16]. The advantages of the quadrupole method are: easily extendable to multiple layers, faster computation time as compared to finite element method simulations, and easy to insert realistically shaped flaws into simulations [17]. For this case, with no convection losses at the vertical or horizontal edges, the heat flux generated is at the flaw location within the structure, as shown in Fig. 15.8. For static loading scenarios, the heat flux is produced during damage formation and has been observed to be instantaneous. A two-layer configuration with a finite lateral dimension is shown in Fig. 15.8 with no heat flow across the vertical edges at $x = 0$ and $x = L$.

The condition of no heat flow across the vertical edges is satisfied if the solution in the x-direction is given by a cosine transform. Each of the coefficients of the cosine transform of the two-dimensional Laplace transform temperature solution can be found by dividing the layer in two layers of thickness *d* and *l-d* each with a matrix form given as:

$$\begin{bmatrix} \tilde{v}_m(d,s) \\ \tilde{f}_m(d,s) - \tilde{f}_m^s/2 \end{bmatrix} = \begin{bmatrix} \cosh[q_m z] & -\frac{1}{K\,q_m}\sinh[q_m z] \\ -K\,q_m \sinh[q_m z] & \cosh[q_m z] \end{bmatrix} \begin{bmatrix} \tilde{v}_m(0,s) \\ \tilde{f}_m(0,s) \end{bmatrix}, \tag{15.2}$$

$$\begin{bmatrix} \tilde{v}_m(l,s) \\ \tilde{f}_m(l,s) \end{bmatrix} = \begin{bmatrix} \cosh[q_m (l-d)] & -\frac{1}{K\,q_m}\sinh[q_m (l-d)] \\ -K\,q_m \sinh[q_m (l-d)] & \cosh[q_m (l-d)] \end{bmatrix} \begin{bmatrix} \tilde{v}_m(d,s) \\ \tilde{f}_m(d,s) + \tilde{f}_m^s/2 \end{bmatrix}, \tag{15.3}$$

$$\text{where} \quad q_m = \sqrt{\frac{s}{\alpha_z} + \frac{\alpha_x}{\alpha_z}\left(\frac{\pi\,m}{L}\right)^2}, \tag{15.4}$$

and $\tilde{v}_m(z,s)$ and $\tilde{f}_m(z,s)$ are the cosine coefficients for the temperature and flux respectively and $\tilde{f}_m^s$ is the cosine coefficient for the source flux [17]. The Laplace transform of the surface temperature and flux are given as $\tilde{v}_m(0,s)$ and $\tilde{f}_m(0,s)$ respectively. In addition, *K* is the composite thermal conductivity, α_z is the thermal diffusivity in the z direction, α_x is the thermal diffusivity in the x direction, *s* is the Laplace complex argument, *l* is the slab thickness, *d* is the depth of the flaw, *m* is the Fourier cosine series coefficients for each lateral index of x, and *L* is the slab width. Given damage at depth $z = d$ and no heat flow at the surface, Eqs. (15.2) and (15.3) can be solved to find the temperature at the surface to be as:

$$\tilde{v}_m(0,s) = \frac{f_m^s \cosh[q_m (l-d)]}{K\,q_m \sinh[q_m l]}. \tag{15.5}$$

Equation (15.5) can be solved by taking into account the discrete positions in *x* by the vector representation:

$$\boldsymbol{V}(0,s) = \boldsymbol{C}(s).\boldsymbol{F}^s(s) \tag{15.6}$$

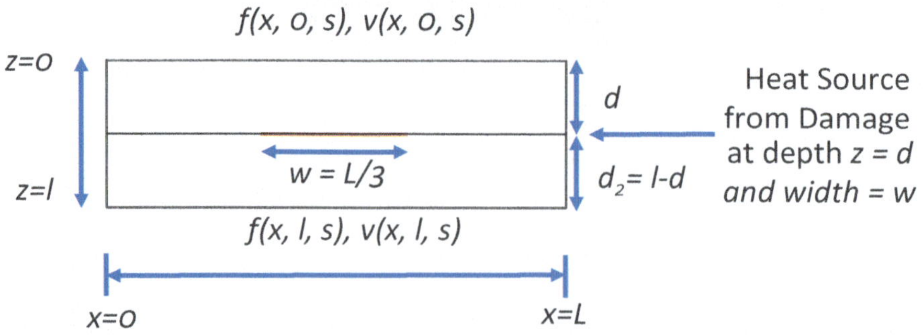

Fig. 15.8 Configuration for two-dimensional model with embedded heat source

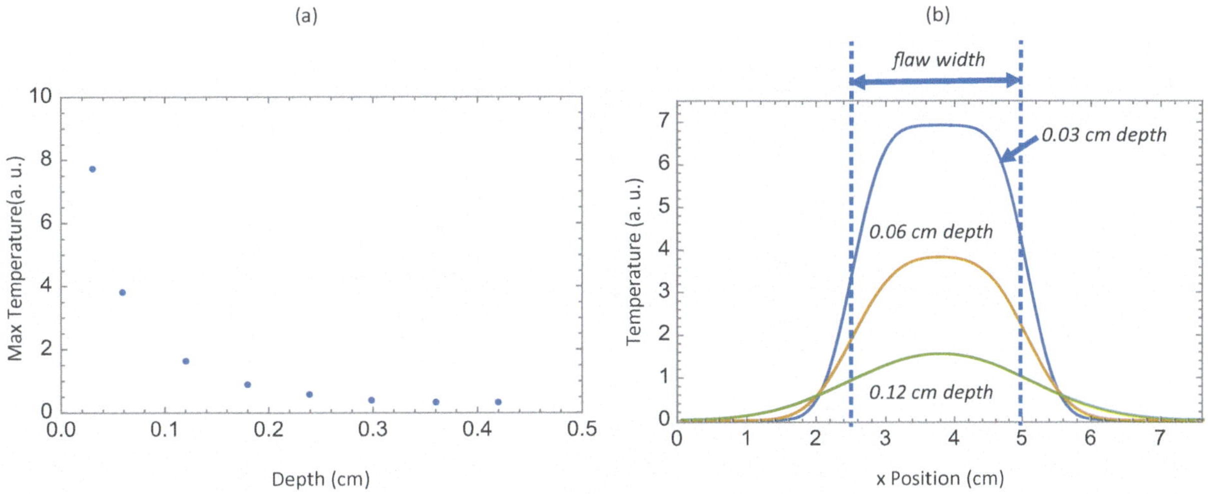

Fig. 15.9 Temperature drop as a function of source depth and blurring of source edge for different depths

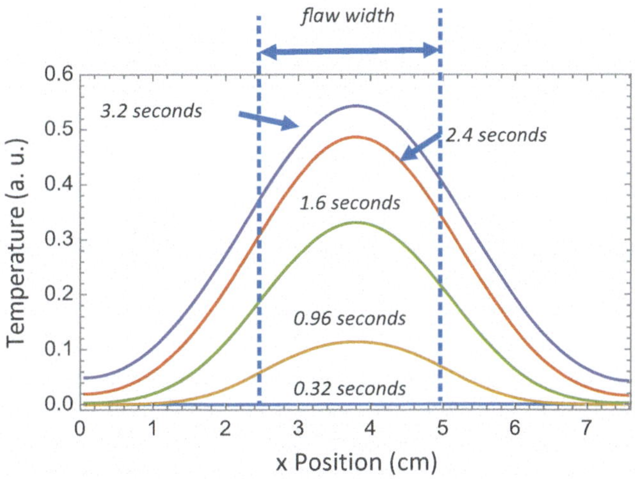

Fig. 15.10 Temperature as a function of time for a flaw source buried 0.24 cm

where the elements of $\boldsymbol{C}(s)$ are given as:

$$C_{i,j}(s) = a_j \frac{\sum_{m=0}^{N-1} a_m \cos\left(\frac{i\,m\,\pi}{N-1}\right) \cos\left(\frac{j\,m\,\pi}{N-1}\right) \cosh\left[q_m\,(l-d)\right]}{(2N-2)\,K\,q_m\,\sinh\left[q_m\,l\right]}. \tag{15.7}$$

Both a_j and a_m are equal to 1 if the index j (or m) = 0 or N-1 and 2 otherwise. The heat flux vector, $\boldsymbol{F}^s(s)$, is a vector representation of the heat generated during the formation of the flaw during loading at depth $z = d$ and is modeled as an impulse function. For each point, the inverse Laplace transform is calculated at a given time to give temperature as a function of x position. The inverse Laplace transform for the surface temperature, $\boldsymbol{V}(0, s)$, is calculated using the Talbot inversion algorithm [18]. Typical results are shown in Figs. 15.9a, b, and 15.10 for maximum temperature as a function of flaw depth, blurring as a function of flaw depth for the maximum temperature and blurring as a function of time for a flaw depth of 0.24 cm respectively. For these plots, the values used were $\alpha_z = 0.00425$ cm^2/s, $\alpha_x = 0.0425$ cm^2/s, l (the slab thickness) = 0.46 cm, $K = 0.00883$ W/cm/K, the $Flux = 1$ Watt/cm^2, flaw width set from 2.5 to 5.0 cm, and L the slab width set to 7.6 cm. As can be seen from Fig. 15.9a, there is considerable temperature drop for sources that are deeper. This is especially problematic if one takes into account camera noise. Additionally, for deep sources, there is considerable blurring of the edges of the source as shown in Fig. 15.9b. There is also considerable blurring over time as shown in Fig. 15.10. This indicates that the earliest detectable thermal signal will give the best sizing of the defect.

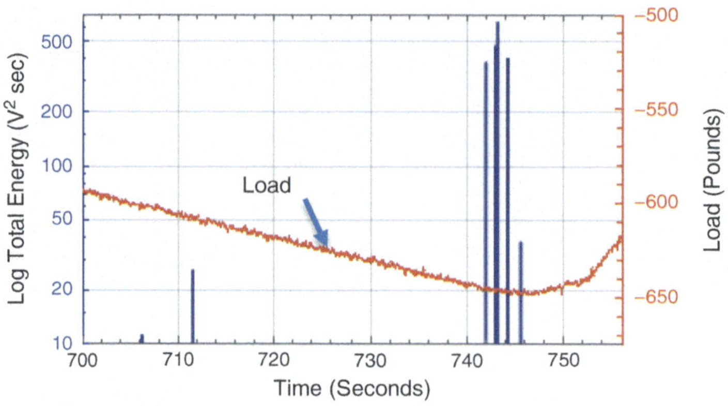

Fig. 15.11 Quasi-static load profile with detected acoustic emission events

15.6 Thermal Imagery Comparison to Acoustic Emission

Acoustic emission data were collected during each quasi-static load test. When a sufficient number of acoustic emission events clustered in a short period of time were detected, this represents rapid damage formation and the loading was stopped to characterize the damage. A typical load profile along with the detected acoustic emission events are shown in Fig. 15.11. The acoustic emission is sensitive to fiber breaks and matrix cracks and these events can produce heat. Shown in Fig. 15.12, are the corresponding processed thermal images. The thermal images show heating resulting from a small edge delamination growing along the edge of the stiffener flange. This is shown in the ultrasound results in Fig. 15.12. This is a successful capture of an early delamination formation that can be grown and characterized during subsequent load runs. The acoustic emission data was compared to thermal data by aligning the recorded load profiles from each system. The results are shown in Table 15.1, for a given acoustic energy event (greater than 10 volts^2 second), the acoustic emission event time is compared to the thermal image indication time. The thermal image indication time is determined from the earliest thermal indication where new heating is observed. This can be seen in Fig. 15.12 at 707.0 s. A small thermal indication is observed and grows as a function of time as the applied load increases. Table 15.1 provides a comparison of the acoustic emission time to the thermal indication time. As shown in Table 15.1 most acoustic emission events corresponded to a detected thermal indication. This shows the sensitivity of the thermal technique to detect and locate the damage. This was helpful since acoustic emission, for these tests and panel setup was not able to determine damage location. For the acoustic emission event at time 711.5 s there was no thermal indication detected. This could be due to the damage located underneath the hat which would be out of the field of view of the infrared camera or a fiber break/matrix crack located at a significant depth that would be undetectable. Nevertheless, thermal imaging provides good agreement with acoustic emission.

The data in Table 15.1 also demonstrates a consistent time delay between the acoustic emission event time and the thermal indication time. This is expected since acoustic emission detection time would be nearly instantaneous as compared to the thermal detection time. The thermal detection time requires the diffusion of heat to the surface. Another way to compare acoustic emission to passive thermography is count the number of pixels above a threshold. This is shown in Fig. 15.13 where the most significant acoustic emission events (energy >100 volts^2 second) are compared to the thermally detected area pixels (pixel values greater than 7 counts shown in dashed line plot) as a function of time. What this clearly shows is the acoustic emission event occurs before a growth in thermally detected damage area. A failure generates an acoustic emission event and heat will take time to diffuse through-the-thickness as well as in-plane.

15.7 Conclusions

It has been shown that passive thermography can be used to detect damage in hat stiffened composite structures during quasi-static loading. The thermally detected damage shape is in good qualitative agreement with the ultrasonic inspection results. In addition, it appears, small thermally detected damage can be detected that are beyond the limits of detection with conventional ultrasound. This is likely a result of the significant loss of ultrasonic signal at the edge of stiffener due to the ultrasonic beam straddling a significant thickness change. The thermal results compared well to the acoustic emission with most acoustic emission events corresponded to a thermal indication thus allowing acoustic emission event location.

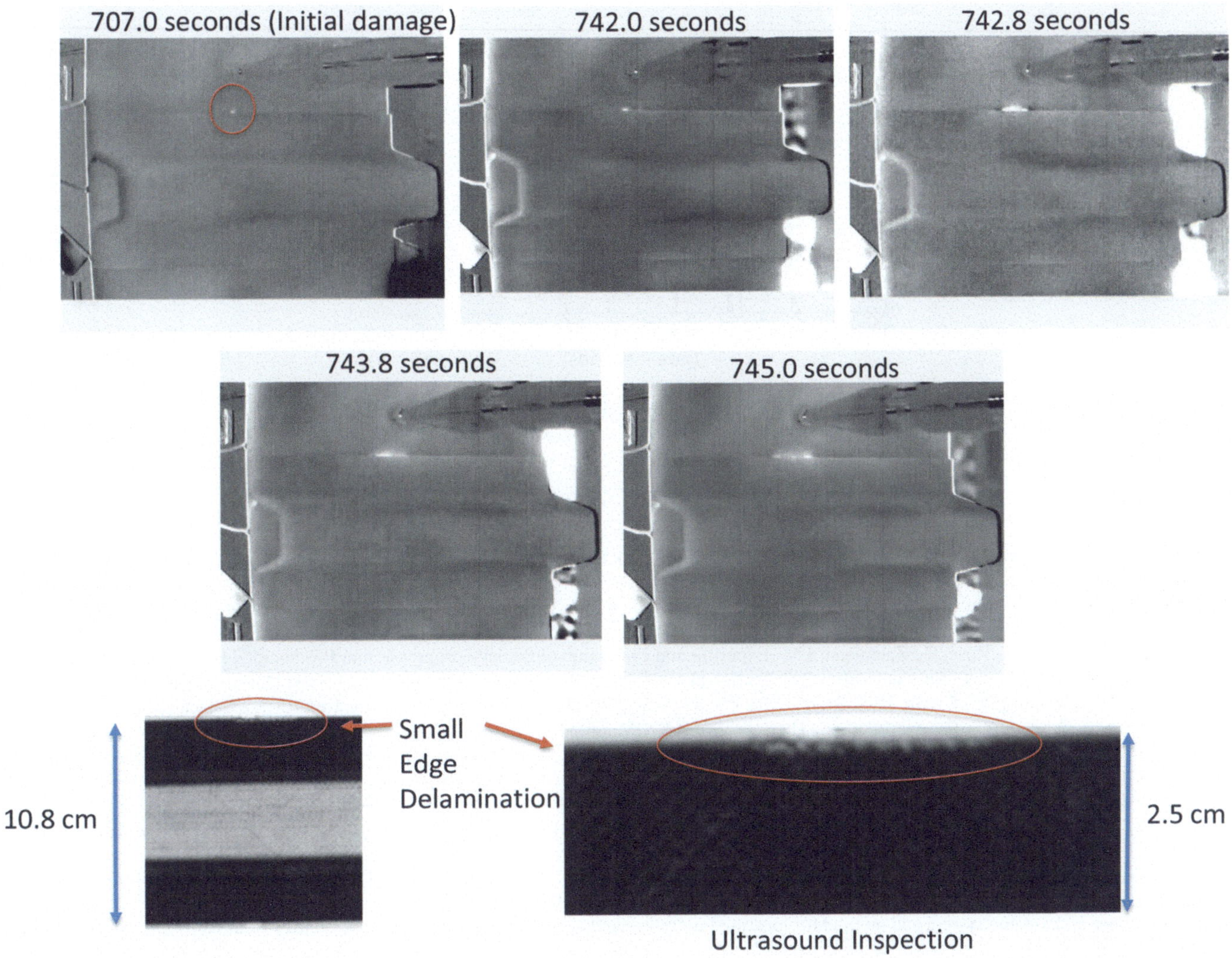

Fig. 15.12 Processed thermal inspection images for a single quasi-static load run with corresponding ultrasound inspection image showing edge delamination

Table 15.1 Comparison of acoustic emission event time to thermal image indication time

Acoustic emission event energy (Volts^2 Second)	Acoustic emission event time (s)	Thermal image indication time (s)
11.4	706.2	707.0
26.5	711.5	Nothing detected
389.7	741.9	742.0
483.7	742.9	743.3
653.6	743.1	744.6
411.0	744.2	745.9
38.3	745.5	747.2

The obvious limitation is the damage must be in the field of view of the thermal imager and at a depth that would allow detection within the detectability limits of the infrared camera. A second thermal camera located on the opposite side (skin side) would be valuable to capture damage that could be hidden by the stiffener hat and could capture the deeper damage facing the opposite side camera. Lastly, a more careful study of the acoustic emission event time compared to the temporal response of the detected thermal transient might provide a start time that would allow fitting of a thermal model to determine damage depth in real-time. This would provide damage location, size and depth.

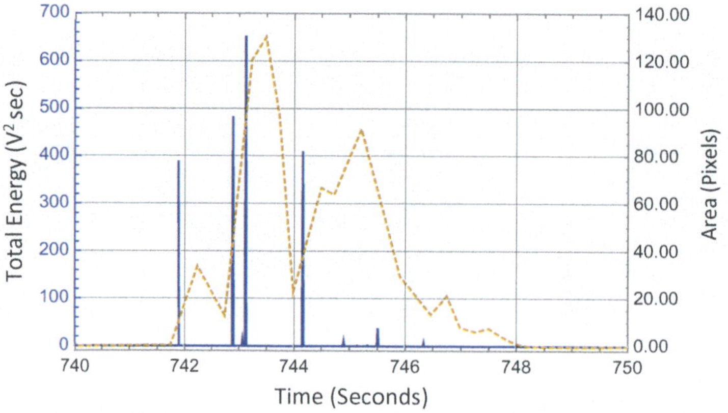

Fig. 15.13 Detected acoustic emission events compared to thermally detected area pixels above a threshold

References

1. O'Brien, T. K.: *Development of a Delamination Fatigue Methodology for Composite Rotorcraft Structure*, NASA Aviation Safety Technical Conference Denver, CO, (October 2008)
2. Krueger, R.: An Approach to Assess Delamination Propagation Simulation Capabilities in Commercial Finite Element Codes, NASA/TM-2008-215123, (2008)
3. Bisagni, C., Dávila, C.G., Rose, C., Zalameda, J.N.: Experimental evaluation of damage progression in Postbuckled single stiffener composite specimens", American Society for Composites 29th Technical Conference Proceedings, US-Japan 16, ASTM D30, (Feb 2014)
4. Zalameda, J.N., Burke, E.R., Parker, R.F., Seebo, J.P., Wright, C.W., Bly, J.B.: Thermography Inspection for Early Detection of Composite Damage in Structures during Fatigue Loading, Thermosense XXXIV, edited by Douglas Burleigh, Gregory R. Stockton, Proc. of SPIE, vol. 835403 (April 2012)
5. La Rosa, G., Clienti, T., Lo Savio, F.: Fatigue analysis by acoustic emission and thermographic techniques, Procedia Engineering, XVII International Colloquium on Mechanical Fatigue of Metals, (ICMFM17), vol. 74, pp. 261–268 (2014)
6. Kordatos, E., Dassios, K., Aggelis, D., Matikas, T.: Rapid evaluation of the fatigue limit in composites using infrared lock-in thermography and acoustic emission. Mech. Res. Commun. **54**, 14–20 (2013)
7. Munoz, V., Vales, B., Perrin, M., Pastor, M.L., Welemane, H., Cantarel, A., Karama, M.: Damage detection in CFRP by coupling acoustic emission and infrared thermography. Compos. Part B. **85**, 68–75 (February 2015)
8. Ringermacher H. I., et al.: System and method for locating failure events in samples under load, United States Patent No. 7,516,663 B2, (April 14, 2009)
9. Zalameda, J.N., Burke, E.R., Horne, M.R., Madaras, E.I.: Large area nondestructive evaluation of a fatigue loaded composite structure, Residual Stress, Thermomechanics & Infrared Imaging, Hybrid Techniques and Inverse Problems, vol. 9, Proceedings of the 2016 Annual Conference on Experimental and Applied Mechanics, ISBN 978-319-42254-1, Chapter 4, (2017)
10. Winfree, W.P., Zalameda, J.N., Howell, P.A.: Improved flaw detection and characterization with difference thermography, Proc. SPIE 8013, 80130U (April 2011)
11. Horne, M.R. Rayleigh wave acoustic emission during crack propagation in steel. PhD Dissertation, Virginia Polytechnic Institute and State University. (2003)
12. Johnston, P.H., Wright, C.W., Zalameda, J.N., Seebo, J.P.: Ultrasonic monitoring of ply crack and delamination formation in composite tube under torsion load, Ultrasonic Symposium (IUS), IEEE, pp. 595–598, (2010)
13. Hourigan, J.S., McIndoo, L.V.: The singular value decomposition. Dec. 1998. College of the Redwoods
14. Zalameda, J.N., Horne, M.R.: Real time detection of damage during quasi-static loading of a single stringer panel using passive thermography, Thermosense XL, edited by Douglas Burleigh, Jaap deVries, to be published Proc. of SPIE (April 2018)
15. Chan, M., (published 2012, March 09), Perspective Control/Correction", Retrieved from URL https://www.mathworks.com/matlabcentral/fileexchange/35531-perspective-control--correction (downloaded Jan. 2018)
16. Winfree, W.P., Zalameda, J.N., Gregory E.D.: Application of the quadrupole method for simulation of passive thermography", Proc. SPIE 10214, Thermosense: Thermal Infrared Applications XXXIX, 102140Y (5 May 2017)
17. Winfree, W.P., Zalameda, J.N., Howell, P.A., Cramer K.E.: Simulation of thermographic responses of delaminations in composites with quadrupole method, Proc. SPIE 9861, Thermosense: Thermal Infrared Applications XXXVIII, 98610N (2016)
18. Talbot, A.: The accurate numerical inversion of Laplace transforms. IMA J. Appl. Math. **23**(1), 97–120 (1979)

Chapter 16
The Effect of Chamber Temperature on Residual Stresses of FDM Parts

C. Casavola, A. Cazzato, D. Karalekas, V. Moramarco, and G. Pappalettera

Abstract The Fused Deposition Modelling (FDM) is nowadays one of the most widespread and employed processes to build complex 3D prototypes directly from a STL model. In this technique, the part is built as a layer-by-layer deposition of a feedstock wire. This typology of deposition has many advantages but it produces rapid heating and cooling cycles of the feedstock material that introduces residual stresses in the part during the building-up. Consequently, warping, de-layering and distortion of the part during the print process are common issues in FDM parts and are related to residual stresses. In the view to reduce this kind of issues, the high-level print systems use a heated chamber. The aim of the present work is to measure the residual stresses in several points of the printed parts, both on top and bottom, in order to verify if the use of the heated chamber during the printing produce substantial variation. The residual stresses have been measured in ABS parts employing the hole-drilling method. In order to avoid the local reinforcement of the strain gage, an optical technique, i.e. ESPI (electronic speckle pattern interferometry), is employed to measure the displacement of the surface due to the stress relaxation and, consequently, to calculate the residual stresses.

Keywords 3D printing · Residual stress · Fused deposition modelling · Hole drilling · ESPI

16.1 Introduction

The Fused Deposition Modelling (FDM) has become one of the most employed techniques to build prototypes and parts directly from a 3D solid model. This technology was invented in the early 1990s by Stratasys and, initially, the design verification, kinematic functionality testing and fabrication of models for visualization were the main fields where it was employed. Nowadays, there are many new fields such as aerospace, medical, construction, cultural [1, 2] and many other where it has been successfully employed. Finally, the diffusion of the low-cost desktop 3D printers such as RepRap, MakerBot, Cube, etc., has made this technology widely accessible even at home and office.

In this process, as for many others 3D printing technologies [3, 4], the model is built as a layer-by-layer deposition of a feedstock material. Initially, the raw material is in the form of a filament that is partially melted, extruded and deposited by a numerically controlled heated nozzle onto the previously built model [1]. After the deposition, the material cools, solidifies and sticks with the surrounding material. Due to the layer-by-layer construction and the orientation of the material deposition, once the entire model has been deposited, the FDM part shows orthotropic material properties with a behaviour similar to a laminate orthotropic structure [5]. Initially, the FDM printers have been able to build parts only in acrylonitrile-butadiene-styrene (ABS) and polylactic acid (PLA). However, nowadays many other materials have been employed and developed, e.g., metals [6], ceramics [7], bio-resorbable polymers (PCL) [8], metal/polymers mixture materials [9], and short fibre composites [10]. The PLA, compared to ABS, have a stronger mechanical resistance and a lower coefficient of thermal expansion. The last property improves the printability of the material because it reduces the de-layering problems and the warp effect during the printing phase. This distortion effect of the part during the printing is one of the most important issues in the FDM process, because it could seriously affect the shape and the final dimensions of the parts or it could prevent the finalization of the objects due to unsticking of the object from the bed. The distortions are due to the continuous rapid heating and cooling cycles of the deposited material [11–14]. A common and cheap technique, in order to reduce this problem, is to employ a heated bed with some type of adhesive on the surface. However, due to bad heat conduction properties of plastic

C. Casavola · A. Cazzato · V. Moramarco (✉) · G. Pappalettera
Dipartimento di Meccanica, Matematica e Management (DMMM) – Politecnico di Bari, Bari, Italy
e-mail: vincenzo.moramarco@poliba.it

D. Karalekas
Laboratory of Advanced Manufacturing Technologies and Testing, University of Piraeus, Piraeus, Greece

A. Baldi et al. (eds.), *Residual Stress, Thermomechanics & Infrared Imaging, Hybrid Techniques and Inverse Problems, Volume 7*,
Conference Proceedings of the Society for Experimental Mechanics Series, https://doi.org/10.1007/978-3-319-95074-7_16

materials, after few deposited layers the effect of the heated bed is lost. Indeed, to tackle this problem in a better way, the expensive and high-level printing systems use a heated chamber. This method, should reduce the problem related to residual stresses and distortion during the printing phase by creating a uniform temperature around the part.

Some papers have dealt with experimental measurements of residual stress distribution in plastic parts [15–17] but few works are about FDM parts [12–14]. Turnbull et al. [15] carried out a comparison among several techniques to measure residual stresses in ABS, Polycarbonate, and Nylon. They concluded that the hole drilling can be employed as a valid measurement method to measure residual stresses in plastic materials. Nau et al. [16] highlighted that the process parameters and procedures applied for stress analysis in metallic materials cannot be employed in polymers. They pointed out that the surface preparation of specimens, the strain gauge bonding, and the drilling speed are critical issues to obtain a correct measure. However, both Turnbull et al. [15] and Nau et al. [16] did not consider the local reinforcement effect that the installation of a rosette produces in materials that have a low Young's modulus. Indeed, Magnier et al. [17] studied the influence of material viscoelasticity, room temperature and local reinforcement of the strain gauge on the measure of deformation by HDM of plastic materials. They highlighted that the use of strain gauge to measure the deformation on plastic materials can produce a difference up to 30% between the results recorded by strain gauges and DIC. Casavola et al. [12–14] studied the effect on residual stresses of the raster orientation in FDM parts. They found that the stacking sequence +45°/−45° shows the lowest values of residual stresses. Moreover, they showed that there is a clear effect of the constraint on the residual stress distribution highlighting that, near the edges of the parts, the residual stresses are higher. Only one paper has tried to deal with the residual stress issues in FDM part by numerical simulation. Zhang and Chou [18], using simplified material properties and boundary conditions, have simulated different deposition patterns and have demonstrated the feasibility of using the element activation function to reproduce the filament deposition. They found that there was a modification of the residual stress distributions if the tool-path pattern is changed. However, they did not validate their model using residual stress measurements but only by comparing the distortion of the printed part and the numerical prediction.

The aim of the present work is to measure the residual stresses in FDM specimens, both on top and bottom, to verify if there are any differences between a 3d printer that has a heated bed and one with the heated chamber. The residual stresses have been measured in ABS parts employing the hole-drilling method. In order to avoid the local reinforcement of the strain gage, an optical technique, i.e. ESPI (electronic speckle pattern interferometry), is employed to measure the displacement of the surface due to the stress relaxation and, consequently, to calculate the residual stresses.

16.2 Materials and Methods

A Stratasys Dimension Elite and a MakerBot Replicator 2X have been employed to produce the specimens (Fig. 16.1). The first one is a professional 3d printer with a heated chamber while the latter is a low-cost desktop 3d printer with a heated bed.

The specimens have a rectangular shape and the dimensions of 80 × 40 × 5 mm with a ±45° stacking sequences. The chosen stacking sequences was due to the limitation of the Stratasys software that does not allow modifying this parameter. Moreover, the samples have been manufactured with the minimum dimension of the part perpendicular to the build platform. The Fig. 16.2 shows the coordinate system for the residual stresses in relation to the specimen shape.

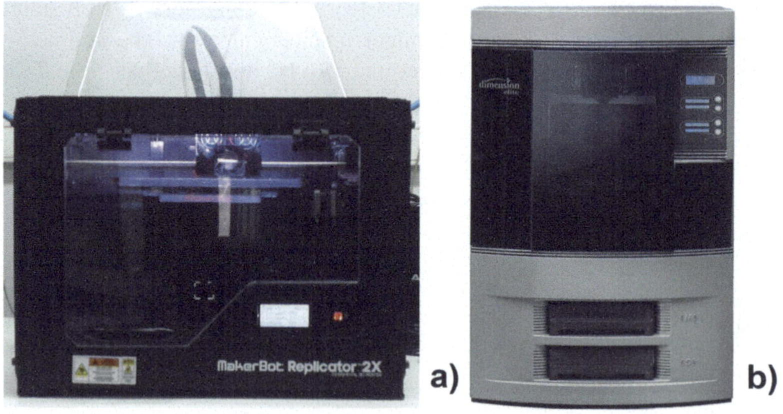

Fig. 16.1 The MakerBot 2X (**a**) and the Stratasys Dimension Elite (**b**) printers

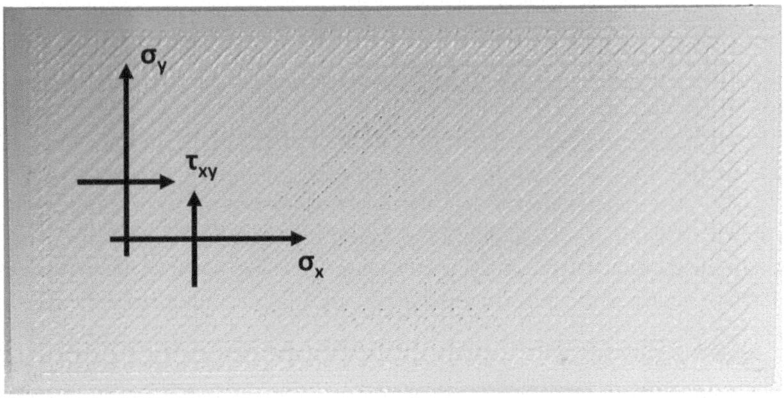

Fig. 16.2 Exemplary specimen with coordinate system for the residual stresses

Table 16.1 Fixed printer parameters

Parameter	Value
Layer thickness [mm]	0.25
Bead width [mm]	0.67
Number of contour lines	2
Bed temperature for MakerBot printer [°C]	110
Temperature of the heated chamber for Stratasys [°C]	75

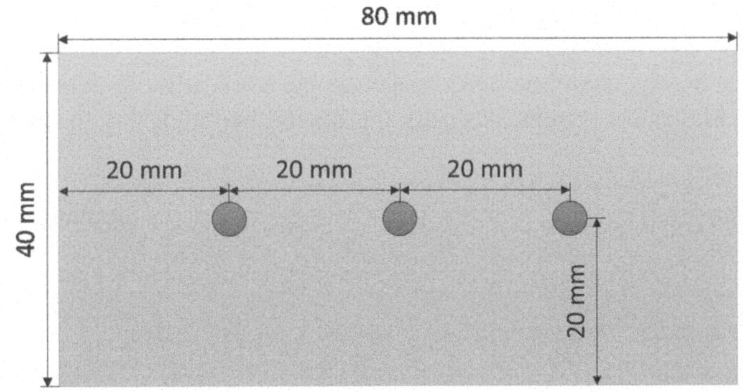

Fig. 16.3 Holes position on the top and the bottom of the specimens

The parameters reported in Table 16.1, such as the layer thickness or the number of contour lines, have been kept constant for every specimen and for each printer.

In Table 16.1, the layer thickness and the bead width are respectively the height and the width of a deposited filament. The number of contours represents how many edges have been deposited before filling the inner part by inclined beads. The solid model, created using a 3D CAD, has been sliced using the proprietary slicing software of each printer.

In this work, the ESPI technique has been employed to measure the displacement around a hole drilled inside the material. Due to the orthotropic behaviour of FDM parts, the isotropic model usually implemented in commercial hole drilling software cannot be used. Thus, an orthotropic FEM model has been developed to calculate the displacements due to some known stress cases. The combination between the experimental displacements data and the FE model allows calculating the residual stress in the parts [13].

In order to calculate the mean values of the residual stresses, the measure has been carried out on three different samples for each printer. Moreover, three holes have been drilled on the top of each specimen, i.e. starting from the last layer deposited, and three on the bottom, i.e. starting from the first layer deposited (Fig. 16.3). Finally, an average value of the three holes has been calculated and the comparison between top and bottom has been carried out.

The holes were drilled by means of a high-speed turbine which is mounted on a precision translational stage. Turbine rotation speed was set to 5000 rpm after some preliminary tests that indicated that this speed allows obtaining good quality holes [11, 19, 20]. The cutter is made by tungsten coated by TiN and it has a nominal diameter d = 1.59 mm. Compressed air was activated during the test to clean the surface of the sample by the formation of drilling chips. The holes were drilled to a depth of 0.6 mm through 12 drill increments to contain the temperature during the drilling phase at lower values.

A diode pumped solid-state laser source ($\lambda = 532$ nm) was used to shine the sample and to generate the speckle pattern. The laser beam is divided in two parts by a beamsplitter and delivered by two optical fibres. The beam emerging out from the first fibre is collimated and then directed towards the sample at a given angle ($\alpha = 53°$). The beam emerging from the second fibre, instead, is directed towards the CCD matrix of the camera and it acts as a reference beam. The CCD camera (640×480 pixel) itself is placed at a given angle with respect to the normal to the sample ($\beta = 40°$). Light diffused by the sample interferes on the CCD matrix with the reference beam. Four-step temporal phase shifting algorithm was adopted in order to obtain the phase [21, 22]. This means that four reference images are taken initially having a $\pi/2$ phase difference among each other. Another set of four images is, analogously, taken for each drill increment. These intensity patterns were subtracted from the reference intensity patterns recorded on the sample before starting the drilling procedure. This operation allows obtaining fringe patterns encoding the information about the displacement experienced by the sample along the sensitivity vector.

16.3 Results and Discussions

The σ_x, σ_y, and τ_{xy} residual stresses trends for the two printers and for the top and the bottom of the specimens have been shown in Figs. 16.4 and 16.5. As pointed out by Casavola et al. [13], the ±45° specimens show quite low residual stresses thought higher than that reported in that work. For the τ_{xy} it cannot be observed any significant difference between the printers and, as pointed out by Casavola et al. [13], the τ_{xy} are quite low (Fig. 16.5). Indeed, the values of residual stress are roughly between −1 MPa and + 1 MPa. The comparison between the MakerBot and the Stratasys residual stresses σ_x and σ_y (Fig. 16.4a, c) show that there is no clear difference on the top of the specimens. On the other hand, there are some differences, between the studied configurations, on the bottom of the specimens (Fig. 16.4b, d). These are remarkable near the surface of the specimens, i.e. until 0.2 mm, while no clear differences can be identified at higher depth [13]. The Stratasys machine, having the heated chamber, can reduce temperature difference between the deposited layer and the previous deposited material. The MakerBot printer, has only the heated bed and, due to the bad heat conduction properties

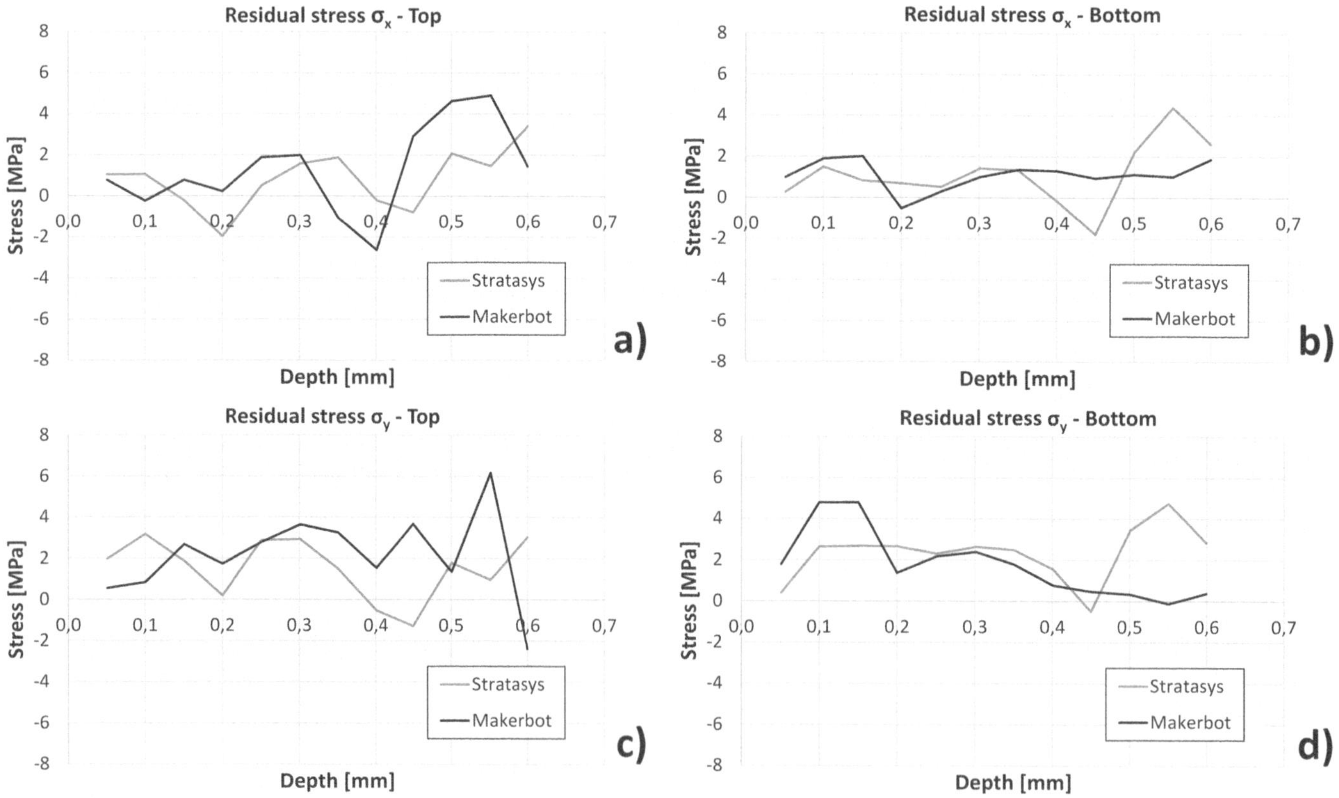

Fig. 16.4 σ_x and σ_y residual stresses for Stratasys and MakerBot printers: (a) σ_x top side; (b) σ_x bottom side; (c) σ_y top side; (d) σ_y bottom side

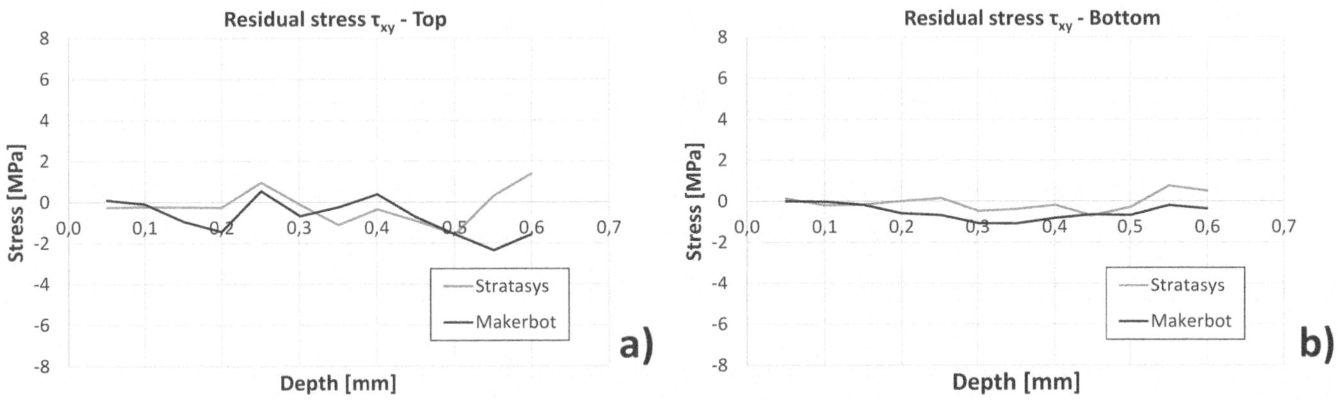

Fig. 16.5 τ_{xy} residual stresses for Stratasys and MakerBot printers: (**a**) top side; (**b**) bottom side

of plastic materials, the beneficial effect of the heated bed is reduced after few deposited layers. This can explain the higher residual stress values of the MakerBot compared to the Stratasys on the bottom of the specimens. However, it should be considered that, the MakerBot is not an open-chamber printer but it has a closed space where the hot-end prints. In this way the chamber temperature, thought is not controlled, is higher than the room temperature due to the heating effect of the bed in a closed space. In conclusion, there is a clear advantage of using a heated chamber to reduce the residual stresses in the printed parts. This difference could be clearer comparing the Stratasys machine with an open chamber machine such as RepRap in which the heat of the bed is dispersed in the room. Furthermore, we have to point out that the chosen stacking sequence ensures the lowest residual stresses, in comparison with others sequences, reducing the advantages of using heated chambers.

16.4 Conclusions

In this paper, the measure of the residual stresses in FDM printed parts has been carried out by the hole-drilling method. In order to avoid the local reinforcement of the strain gage, an optical technique, i.e. ESPI (electronic speckle pattern interferometry), is employed to measure the displacement of the surface due to the stress relaxation and, consequently, calculate the residual stresses. The aim of the work was to measure the residual stresses in several points of 3D printed parts, both on top and bottom, in order to verify if there is a clear difference between a heated bed machine and a heated chamber one. This comparison has been carried out between two printers a Stratasys Dimension Elite and a MakerBot Replicator 2X.

The results show that, whereas for σ_x and σ_y there are some differences in the bottom of the specimens, for σ_x and σ_y on the top of the specimens and τ_{xy} it cannot be observed any difference between the two printers. The differences are remarkable near the surface of the specimens, i.e. until 0.2 mm, but below this depth, there are no clear differences between the studied configurations. The Stratasys machine, having the heated chamber, can reduce temperature difference between the deposited layer and the previous deposited material and in this way the residual stress values are reduced if compared to the MakerBot. However, it should be considered that the MakerBot is not an open-chamber printer but it has a closed space where the hot-end prints. In this way the chamber temperature, thought is not controlled, is higher than the room temperature due to the heating effect of the bed in a closed space. In conclusion, there is an advantage of using a heated chamber to reduce the residual stresses in the printed parts. This difference could be clearer comparing the Stratasys machine with an open chamber machine such as RepRap in which the heat of the bed is dispersed in the room.

References

1. Yan, X., Gu, P.: A review of rapid prototyping technologies and systems. Comput. Aided Des. **28**(4), 307–318 (1996)
2. Petzold, R., Zeilhofer, H.F., Kalender, W.A.: Rapid prototyping technology in medicine—basics and applications. Comput. Med. Imaging Graph. **23**(5), 277–284 (1999)
3. Chua, C.K., Chou, S.M., Wong, T.S.: A study of the state-of-the-art rapid prototyping technologies. Int. J. Adv. Manuf. Technol. **14**(2), 146–152 (1998)

4. Barile, C., et al.: Mechanical characterization of SLM specimens with speckle interferometry and numerical optimization. Conf. Proc. Soc. Exp. Mech. Ser. **6**, 837–843 (2011)
5. Casavola, C., et al.: Orthotropic mechanical properties of fused deposition modelling parts described by classical laminate theory. Mater. Des. **90**, 453–458 (2016)
6. Mireles, J., et al.: Development of a fused deposition modeling system for low melting temperature metal alloys. J. Electron. Packag. **135**(1), 011008–011008 (2013)
7. Allahverdi, M., et al.: Processing of advanced electroceramic components by fused deposition technique. J. Eur. Ceram. Soc. **21**(10–11), 1485–1490 (2001)
8. Zein, I., et al.: Fused deposition modeling of novel scaffold architectures for tissue engineering applications. Biomaterials. **23**(4), 1169–1185 (2002)
9. Masood, S.H., Song, W.Q.: Development of new metal/polymer materials for rapid tooling using fused deposition modelling. Mater. Des. **25**(7), 587–594 (2004)
10. Zhong, W., et al.: Short fiber reinforced composites for fused deposition modeling. Mater. Sci. Eng. A. **301**(2), 125–130 (2001)
11. Kantaros, A., Karalekas, D.: Fiber Bragg grating based investigation of residual strains in ABS parts fabricated by fused deposition modeling process. Mater. Des. **50**, 44–50 (2013)
12. Casavola, C., et al.: Preliminary study on residual stress in FDM parts. In: Residual Stress, Thermomechanics & Infrared Imaging, Hybrid Techniques and Inverse Problems, vol. 9, pp. 91–96. Springer International Publishing, New York LLC (2017)
13. Casavola, C., et al.: Residual stress measurement in fused deposition modelling parts. Polym. Test. **58**, 249–255 (2017)
14. Casavola, C., et al.: Influence of printing constraints on residual stresses of FDM parts. Conf. Proc. Soc. Exp. Mech. Ser. **8**, 121–127 (2018)
15. Turnbull, A., Maxwell, A.S., Pillai, S.: Residual stress in polymers - evaluation of measurement techniques. J. Mater. Sci. **34**(3), 451–459 (1999)
16. Nau, A., et al.: Application of the hole drilling method for residual stress analyses in components made of polycarbonate. Zeitschrift Kunststofftechnik/J. Plastics Technol. **2011**(3), 66–85 (2011)
17. Magnier, A., A. Nau, B. Scholtes. Some aspects of the application of the hole drilling method on plastic materials, in Conference Proceedings of the Society for Experimental Mechanics Series, (2016)
18. Zhang, Y., Chou, Y.: Three-dimensional finite element analysis simulations of the fused deposition modelling process. Proc. Inst. Mech. Eng. B J. Eng. Manuf. **220**(10), 1663–1671 (2006)
19. Barile, C., et al.: Drilling speed effects on accuracy of HD residual stress measurements. Conf. Proc. Soc. Exp. Mech. Ser. **8**, 119–125 (2014)
20. Barile, C., et al.: Residual stress measurement by electronic speckle pattern interferometry: a study of the influence of geometrical parameters. Struct. Integrity Life. **11**(3), 177–182 (2011)
21. Kujawinska, M.: Use of phase-stepping automatic fringe analysis in moire interferometry. Appl. Opt. **26**(22), 4712–4714 (1987)
22. Ghiglia, D.C., Pritt, M.D.: Two-Dimensional Phase Unwrapping: Theory, Algorithms, and Software, vol. 4. Wiley, New York (1998)

Chapter 17
System Identification of Structures with Incomplete Modal Information

Chang-Sheng Lin and Ming-Hsien Lin

Abstract The topic of this paper is system identification of structures conducted with consideration of incomplete modal information. A continuum structure has infinite number of degrees of freedom and an infinite number of modes. Modal estimation method is therefore not employed to identify all modes of systems. In addition, incomplete modal information may be led from insufficient response data in practical vibration measurement. In this paper, by using channel-expansion technique, modal estimation from response data of insufficient channel can be performed through the Ibrahim time-domain method. Applicability and effectiveness of the proposed method is demonstrated by numerical simulation of a chain model. Identification of the mode shapes, however, is still a challenging problem to be resolve in the proposed method from incomplete modal information data.

Keywords Incomplete modal information · Modal estimation · Channel-expansion technique · Ibrahim time-domain method

17.1 Introduction

When performing modal analysis of structures based on the theory of system identification, the finite element model is usually constructed to demonstrate a realistic structural system. Due to the accuracy requirements of a practical complex structure itself, the high order (degree of freedom, DOF) of the finite element model is usually produced accordingly. However, in the practical vibration measurement, due to the economic restrictions and structural unavailability, i.e., measurement location within the structure, and measurement technology is immature, then the sensors placed on overall DOFs of a structure to measure response data is not available. Therefore, the number of degrees of freedom corresponding to a mode shape vector to be identified and the order of finite element models are inconsistent, so that it causes the problems of incomplete number of modes and incomplete degree of freedom of mode shape vector.

During the 1970s, Ibrahim proposed a method developed in the time domain, which is usually referred to as the Ibrahim time domain method (ITD Method) [1], and is applied to problems involving the free-decay response data of structures, and is employed to perform modal identification via an eigenvalue analysis. Based on the Prony's theory, Brown et al. developed the least square complex exponential algorithm (LSCE) [2] using a squared output matrix constructed by multichannel impulse response functions. The pseudo-inverse technique is employed to estimate the coefficients of the Prony's polynomials and then extract the modal parameters of a system through the Prony's technique. In 1982, Vold et al. further proposed poly reference complex exponential method (PRCE) [3] to perform modal identification for the case that one of the modes may not be present in the response data. In 1985, among follow-up developments on minimal realization algorithm and singular value decomposition (SVD) [4], Juang and Pappa [5] proposed the Eigensystem Realization Algorithm (ERA) using the impulse response or the free-vibration response of the system to construct the Hankel matrix, which is an augmented matrix containing Markov parameters, for reducing the effect of noise, and making the parameters estimation more accurate.

C.-S. Lin (✉) · M.-H. Lin
Department of Vehicle Engineering, National Pingtung University of Science and Technology, Pingtung County, Taiwan
e-mail: changsheng@mail.npust.edu.tw

A. Baldi et al. (eds.), *Residual Stress, Thermomechanics & Infrared Imaging, Hybrid Techniques and Inverse Problems, Volume 7*, Conference Proceedings of the Society for Experimental Mechanics Series, https://doi.org/10.1007/978-3-319-95074-7_17

17.2 ITD Method and Channel Expansion Technique

The ITD method is based on the state equations of a dynamic system. A simplified version of the method is presented here. For simplicity of explanation, assume that the structure is measured in n places (this assumption can be relaxed) at 2n times, where n is the number of degrees of freedom exhibited by the test structure. From the measured free-decay responses at n stations on a structure under test, each with q sampling points, we define a system matrix [**A**], which is an $n \times n$ matrix, such that

$$[\mathbf{A}]\,[\mathbf{X}] = [\mathbf{Y}] \tag{17.1}$$

where [**X**] and [**Y**] are, respectively, $n \times q$ data-expansion matrices of the free decay and its time-shifted response. The number q is generally chosen to be larger than the number of measurement channels n, the system matrix [**A**] can be therefore estimated from [**X**] and [**Y**] through the least-squares method. Once the system matrix [**A**] is obtained via least-squares analysis from measured data, the modal parameters of the structural system can be determined by solving the eigenvalue problem associated with the system matrix [**A**]. Brief speaking, ITD method, which is the same as the most time domain approaches, constructs a matrix from the time response and numerically computes the modal data by solving an eigenvalue problem.

In reality, we do not know in advance how many modes are required to describe the dynamic behavior of the observed structural system. The number of (real) modes m involved in the response determines the number of measurement channels, which is chosen to be at least twice of the number of modes of interest to appropriately identify the $2m$ complex modes. If the number of measurement channels does not actually reach $2m$, we may employ the technique of channel expansion [3] with sampling time shifted to reach the total available number of measurement channels. It should be noted that, however, the identified mode shapes are composed of the components corresponding only to those physically measured response channels. In addition, due to the fact that the results of modal identification may be poor from the noise effect, through the channel-expansion technique in ITD method, which uses time-delayed sampling points from the original response to increase the total numbers of sampling points and measurement channels, we can therefore reduce the effect of noise to improve the accuracy of modal estimation based on the property of consistency in the theory of system identification.

17.3 Numerical Simulations

Consider a 6-dof system of the chain model with viscous damping, whose mass matrix $\boldsymbol{M}$, stiffness matrix $\boldsymbol{K}$, and damping matrix $\boldsymbol{C}$ of the system are given as follows:

$$\boldsymbol{M} = \begin{bmatrix} 2&0&0&0&0&0\\ 0&2&0&0&0&0\\ 0&0&2&0&0&0\\ 0&0&0&2&0&0\\ 0&0&0&0&3&0\\ 0&0&0&0&0&4 \end{bmatrix} N\cdot s^2/m \quad \boldsymbol{K} = 600\cdot \begin{bmatrix} 1&-1&0&0&0&0\\ -1&2&-1&0&0&0\\ 0&-1&2&-1&0&0\\ 0&0&-1&2&-1&0\\ 0&0&0&-1&3&-2\\ 0&0&0&0&-2&5 \end{bmatrix} N/m$$

$$\boldsymbol{C} = 0.05\boldsymbol{M} + 0.001\boldsymbol{K}\; N\cdot s/m$$

Note that the system has proportional damping, because the damping matrix $\boldsymbol{C}$ can be expressed as a linear combination of $\boldsymbol{M}$ and $\boldsymbol{K}$. The simulated impulse function serves as the excitation input acting on the sixth mass point of the system. Assume the system is initially at rest, and the displacement responses of the system can be obtained using Newmark's method. The results of modal identification are summarized in Tables 17.1, 17.2, 17.3 and 17.4, which shows that the errors in both natural frequencies and damping ratios are less than 1%, whether using the complete and incomplete response data from the corresponding overall or part of degree of freedom of this system, respectively. However, it is good agreement between the identified and exact mode shapes only from the complete response data only from the corresponding overall degree of freedom of this system. It may be because that the identified mode shapes are composed of the components corresponding only to those physically measured response channels, the pseudo response data through the channel-expansion technique is not available in identification of mode shape.

Table 17.1 Results of modal identification of a 6-DOF system from complete impulse response of overall degree of freedom

	Natural frequency (rad/s)			Damping ratio (%)		
Mode	Exact	ITD	Error (%)	Exact	ITD	Error (%)
1	5.03	5.03	0.00	1.25	1.25	0.00
2	13.44	13.44	0.00	1.04	1.04	0.00
3	19.79	19.79	0.00	1.24	1.24	0.00
4	26.68	26.67	0.00	1.52	1.52	0.00
5	31.65	31.64	0.00	1.74	1.74	0.00
6	33.72	33.71	0.00	1.83	1.83	0.00

Table 17.2 Results of modal identification of a 6-DOF system from incomplete impulse response of 2nd~6th degree of freedom

	Natural frequency (rad/s)			Damping ratio (%)		
Mode	Exact	ITD	Exact	Exact	ITD	Error (%)
1	5.03	5.033	0.00000	1.25	1.245	0.00006
2	13.44	13.45	0.00001	1.04	1.044	0.00032
3	19.79	19.80	0.00001	1.24	1.242	0.00058
4	26.68	26.68	0.00001	1.52	1.521	0.00095
5	31.65	31.65	0.00018	1.74	1.741	0.00363
6	33.72	33.72	0.00003	1.83	1.834	0.00060

Table 17.3 Results of modal identification of a 6-DOF system from incomplete impulse response of 1st~3rd degree of freedom

	Natural frequency (rad/s)			Damping ratio (%)		
Mode	Exact	ITD	Exact	Exact	ITD	Error (%)
1	5.03	5.03	0.00	1.25	1.25	0.00
2	13.44	13.45	0.00	1.04	1.04	0.00
3	19.79	19.80	0.00	1.24	1.24	0.01
4	26.68	26.68	0.01	1.52	1.52	0.01
5	31.65	31.66	0.01	1.74	1.74	0.02
6	33.72	33.72	0.01	1.83	1.83	0.02

Table 17.4 Results of modal identification of a 6-DOF system from incomplete impulse response of 4th~6th degree of freedom

	Natural frequency (rad/s)			Damping ratio (%)		
Mode	Exact	ITD	Exact	Exact	ITD	ITD
1	5.03	5.03	0.00	1.25	1.25	0.102
2	13.44	13.45	0.00	1.04	1.04	0.002
3	19.79	19.80	0.00	1.24	1.24	0.002
4	26.68	26.68	0.01	1.52	1.52	0.002
5	31.65	31.66	0.01	1.74	1.74	0.003
6	33.72	33.72	0.01	1.83	1.83	0.006

17.4 Conclusions

This paper mainly investigates the influence of incomplete vibration measurement information on modal estimation of structures. Through the channel- expansion technique using the time-delayed sampling response data, the Ibraham time domain method is used for modal estimation from response data with incomplete measurement information of structures. Numerical simulation shows that the proposed method can be employed to effectively identify the natural frequency and damping ratio of a chain model system, but identification of mode shapes is still a challenging problem to be resolved in the proposed method.

References

1. Ibrahim, S.R., Mikulcik, E.C.: The experimental determination of vibration parameters from time responses. Shock Vibration Bull. **46**(Part 5), 183–198 (1976)
2. Brown, D.L., Allemang, R.L., Zimmerman, R.D., Mergeay, M.: Parameter estimation techniques for modal analysis, SAE Technical Paper 790221, (1979)
3. Vold, H., Rocklin, G.F.: The numerical implementation of a multi-input modal estimation method for mini-computers, International Modal Analysis Conference Proceedings, (1982)
4. Juang, J.N., Pappa, R.S.: An Eigensystem realization algorithm for modal parameter identification and modal reduction. J. Guid. Control Dyn. AIAA. **8**(5), 620–627 (1985)
5. Juang, J.N., Pappa, R.S.: Effects of noise on modal parameters identified by the Eigensystem realization algorithm. J. Guid. Control Dyn. AIAA. **9**(3), 294–303 (1986)

Chapter 18
Towards Integrating Imaging Techniques to Assess Manufacturing Features and In-Service Damage in Composite Components

I. Jiménez-Fortunato, D. J. Bull, J. M. Dulieu-Barton, and O. T. Thomsen

Abstract Initial work on combining Thermoelastic Stress Analysis (TSA) and Digital Image Correlation (DIC) is presented with the purpose of exploring their used in assessing defects in composite materials. TSA is usually performed using photon detector cameras that are expensive, a further objective of the paper is to investigate the capabilities of low-cost bolometer IR cameras for TSA. A carbon fibre reinforced polymer (CFRP) sample with artificial 'waviness' was created, so that a crack was developed inside the material. Both TSA and lock-in DIC methods detected the damage inside the CFRP specimen. Using the bolometer for TSA has highlighted the need for longer data capture times and the deleterious effects of other features associated with image capture that are inherent in the bolometer system.

Keywords Lock-in · DIC · TSA · Composite materials · Complex structures

18.1 Methodology

The TSA technique is based on the thermoelastic effect which is the reversible relationship between mechanical deformation and thermal energy in an elastic solid when occurring under adiabatic conditions. TSA requires a cyclic load to perform the lock-in processing necessary to extract the small temperature change that is related to the sum of the change in the principal stresses [1]:

$$\Delta T = -\frac{T_0}{\rho C_p}(\alpha_1 \Delta\sigma_1 + \alpha_2 \Delta\sigma_2) \tag{18.1}$$

where ΔT is the temperature change, α is the coefficient of linear thermal expansion, ρ is the material density, C_p is the specific heat capacity at constant pressure and $\Delta\sigma$ are the change in stresses in the principal directions 1 and 2.

Commonly, for TSA, photon detector cameras are used due to their instantaneous response to temperature changes. A photon detector works by counting the photons of energy within the infrared spectrum (typically 3–5 μm) that strike the sensor elements on a cooled detector array and are converted into electrons that are stored in an integration capacitor. The sensors are electronically shuttered by controlling the integration time, i.e. the amount of time the capacitor can collect electrons. The photon detector used in this paper is a FLIR SC5000 and has a near instantaneous response time of ~5.51 μs. However, due to the high capital costs of the equipment and bulkiness of the hardware a low-cost bolometer IR camera is considered as an alternative approach. Bolometer cameras work by transforming the absorbed electromagnetic radiation into electrical signal by using the electrical resistance of the sensor material [2], in this case the bolometer used is a Vanadium Oxide (VOx) semiconductor (FLIR A655sc). However, a key limitation of the bolometer camera is its longer response time ~24 ms which for fast changes in temperature pose a significant challenge, particularly when observing rapidly changing temperature fields such as those generated when performing TSA, where high loading frequencies are require to attain adiabatic conditions.

I. Jiménez-Fortunato (✉) · D. J. Bull · O. T. Thomsen
Infrastructure Research Group, Faculty of Engineering and the Environment, University of Southampton, Southampton, UK
e-mail: ijf1a17@soton.ac.uk

J. M. Dulieu-Barton
Engineering Materials Research Group, Faculty of Engineering and the Environment, University of Southampton, Southampton, UK

A. Baldi et al. (eds.), *Residual Stress, Thermomechanics & Infrared Imaging, Hybrid Techniques and Inverse Problems, Volume 7*, Conference Proceedings of the Society for Experimental Mechanics Series, https://doi.org/10.1007/978-3-319-95074-7_18

The LIDIC technique [3] enables DIC to be performed on a component subject to cyclic loading without the need to synchronise the image capture with the maximum and minimum loads in the cycle. The process used is identical to the lock-in process used in TSA, which by capturing many images allows the amplitude and the phase to be extracted from a noisy sinusoidal signal. In LIDIC, it is not necessary to use high speed cameras, as the data processing allows under-sampling and reconstruction at an apparent lower frequency enabling the lock-in processing to take place. The shortcoming is that sample times are longer and care must be taken to ensure damage is not progressing during the sampling time. By adopting the lock-in process for the DIC it is possible to combine TSA and DIC to obtain a stress metric from the TSA (see Eq. (18.1)) and the component strain changes with the LIDIC. As an example, Chen et al. [4]. carried out crack growth analysis in welded and non-welded T-joints by combining TSA and LIDIC.

18.2 Results and Discussion

As an initial experiment, a 150 mm long × 35.81 mm in width and 5.16 mm thick CFRP $[0,90]_{6s}$ specimen containing artificial waviness and a developed crack, was considered. The specimen was cyclically loaded at 25 kN mean and 20 kN amplitude at 5 Hz. TSA and LIDIC were performed, using frame rates of 383 Hz, 50 Hz and 0.975 Hz for the photon detector, the bolometer and the white light cameras respectively.

The LIDIC results (Fig. 18.1a) shows a strain concentrated region indicating the presence of a crack with a length of ~22 cm. The strain level of the crack is ~0.6% and far field strains are ~0.3%–0.4%. TSA results were obtained with the photon detector (Fig. 18.1b) and the bolometer (Fig. 18.1c). Both IR cameras were able to detect the crack by showing an increased temperature change at the vicinity of the crack, in this case ~0.1 K for the photon detector but a lower response of ~0.04 K for the bolometer.

The difference in response between the photon detector and bolometer cameras is due to the thermal time constant, which is a parameter fixed by the material used for the sensor in the bolometer camera. For the FLIR A655sc camera, the thermal time constant is 8 ms. It has been demonstrated that the time required for the detector to respond fully is at least 3 times the thermal time constant [5], i.e. 24 ms for the FLIR A655sc. The influence of the thermal time constant was studied by performing experiments at different loading frequencies and frame rates on a 316L stainless steel specimen. As the thermoelastic constant of the material is known [4] it was possible to load the specimen with cyclic loads to achieve a ΔT of 100 mK. Fig. 18.2 shows how the reduction in loading frequency allows the bolometer achieves the target temperature change; changing the camera framing rate appears to have a little effect on the response time. Furthermore, Table 18.1 shows how fast the temperature is changing at each loading frequency. The calculation has been performed by considering that

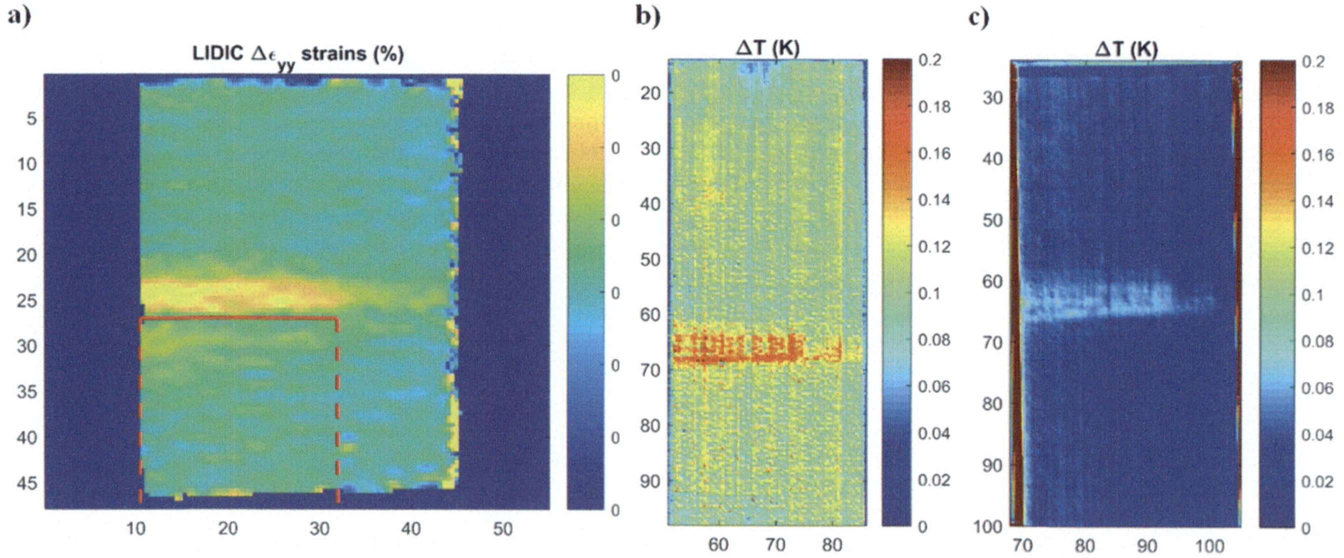

Fig. 18.1 (**a**) Stereo LIDIC, (**b**) TSA with the photon detector, (**c**) TSA with the bolometer

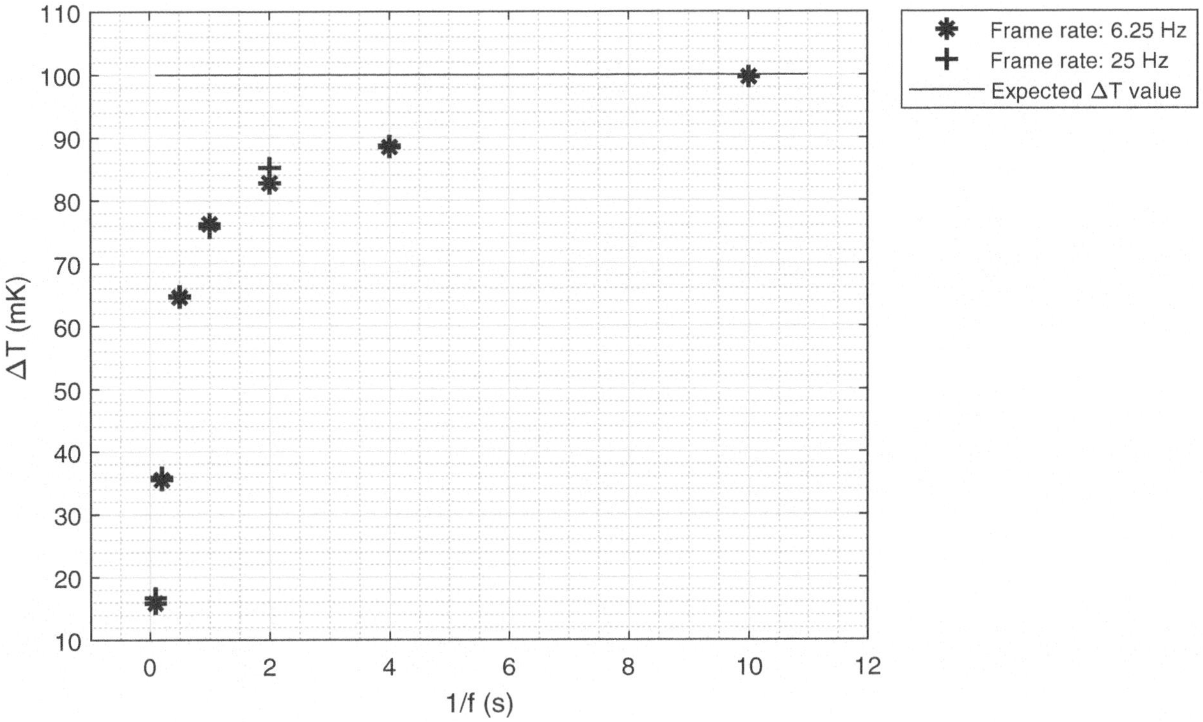

Fig. 18.2 ΔT (mK) versus loading period (s)

Table 18.1 Effect of Loading Frequency on measured value of ΔT

		Loading frequency (Hz)						
ΔT/(1/f/2) in K/s		**11**	**5**	**2**	**1**	**0.5**	**0.25**	**0.1**
Sampling frequency (Hz)	**50**	–	–	–	0.1218	0.0787	0.0443	–
	25	0.3672	0.3590	0.2591	0.1514	0.0852	0.0443	–
	12.5	0.3718	0.3526	0.2576	0.1514	0.0852	0.0443	–
	6.25	0.3489	0.3555	0.2584	0.1525	0.0827	0.0442	0.0199
	3.13	0.0306	0.0243	0.0205	0.1495	0.0550	0.0397	–
Expected ΔT/(1/f/2) in K/s		*2.2000*	*1.0000*	*0.4000*	*0.2000*	*0.1000*	*0.0500*	*0.0200*

the temperature change, which is obtained by the difference between the maximum and the minimum peak temperatures in a cycle, occurs every half loading period. However, if the expected temperature change (0.1 K) is divided by half loading period (assuming a triangular wave), it can be seen that only the 0.1 Hz loading frequency gives a rate of temperature change that can be captured within a frame and there is considerable mismatch for the 11 Hz, 5 Hz and 2 Hz loading frequencies, showing that the bolometer only responds quickly enough for TSA at high loading frequencies.

18.3 Conclusions and Future Work

It was demonstrated that TSA and LIDIC detected damage on the CFRP specimen. However, differences in response between the photon detector and the bolometer were noticed, such as the need for the bolometer to record at lower loading frequency to get a precise measurement. To make the bolometer perform satisfactorily, signal processing approaches, e.g. [6], are required to account for the thermal time constant and this will be an area of focus in future work. Moreover, the combination of TSA and LIDIC will be investigated to get damage characterisation based on the stress (TSA) and strain (LIDIC) fields [4].

References

1. Rajic, N., Street, N.: A performance comparison between cooled and uncooled infrared detectors for thermoelastic stress analysis. Quant. Infrared Thermography J. **11**(2), 207–221 (2014)
2. Gade, R., Moeslund, T.B.: Thermal cameras and applications: a survey. Mach. Vis. Appl. **25**, 245–262 (2014)
3. Fruehmann, R.K., et al.: The use of a lock-in amplifier to apply digital image correlation to cyclically loaded components. Opt. Lasers Eng. **68**, 149–159 (2015)
4. Chen, D., et al.: Crack growth analysis in welded and non-welded T-joints based on lock-in digital image correlation and thermoelastic stress analysis. Int. J. Fatigue. **110**, 172–185 (2018)
5. Budzier, H., et al.: Fast microbolometer-based infrared camera system, presented at the 9th International Conference on Infrared Sensors & Systems, Nürnberg, Germany (2006)
6. Rajic, N., Rowlands, D.: Thermoelastic stress analysis with a compact low-cost microbolometer system. Quant. Infrared Thermography J. **10**(2), 135–158 (2013)

Chapter 19
Image-Based Stress Field Reconstruction in Complex Media

R. Seghir, F. Pierron, and L. Fletcher

Abstract In many instances in life, materials are subject to deformation at high rates, for example: impact, crash, metal forming or pulsed welding. In this context, the transient and inhomogeneous nature of such loading as well as the strong multi-physic couplings induced by quasi-adiabatic conditions make: the experimental capture of the mechanical response very challenging. Additionally, assumptions regarding the constitutive relation of the deforming material are generally required. To overcome both issues, we demonstrate that experimental full-field measurements of acceleration fields can be directly used to invert the local equilibrium equation and reconstruct fields of the stress tensor with no assumption on the constitutive relation and its spatial and temporal variations. We also demonstrate that both experimental stress and strain fields can be recombined to eventually identify the local tangent stiffness tensor of the material. This study constitutes a first step in the field of "direct model identification", as opposed to standard parametric model identification.

Keywords Ultra-high speed imaging · Dynamic · Stress · Model identification · Mechanical behaviour

19.1 Introduction

When localised deformation occurs in a material due to the loading *e.g.* dynamic waves, to instabilities *e.g.* necking, adiabatic shear-bands or simply when cracking, assumptions or models are required to connect experimental measurements at the VER scale to the true local (thermo-) mechanical state. In the last 30 years there has been a massive expansion in the use of full-field measurement techniques for model validation. However, such models still cannot be validated at the scale they are formulated since local stresses are still inaccessible. Furthermore, in many cases only energy quantities are sought as in fracture mechanics where the fracture energy could be simply derived from a local energy balance, regardless of the constitutive equation, if only the stress and strain fields were accessible. Recent developments in image-based techniques have shown that acceleration fields can be experimentally captured during an impact event [1]. Using ultra-high speed imaging (greater than 1 MHz), the acceleration fields can be used as an embedded load cell if the material density is known [2]. This allows for a contactless measurement of both averaged strain and stress in the microsecond regime. More recently, such a methodology has allowed for the capture of a variety of local dynamic stress-strain responses within PMMA specimens subjected to unidirectional gradients of temperature and strain-rate [3]. This has confirmed the ability of the image-based methods to quantitatively capture the local mechanical response under a 1D loading condition. In this paper, we extend this concept to full-field stress tensor reconstruction from the acceleration fields. This new methodology is termed image-based stress reconstruction (IBSR). Explicit dynamic simulations are used to validate this methodology. Using simulation data, each component of the stress field is recovered from acceleration fields under plane stress assumptions. The material stiffness maps are then back calculated by recombining the "measured" strain and the identified stress. While only elastic homogeneous simulations are used in the following, the methodology is generic and independent of the material behavior. In the future this methodology will be applied to experimental data for full-field stress reconstruction.

R. Seghir (✉)
Research Institute in Civil and Mechanical Engineering (GeM), Ecole Centrale de Nantes, Université de Nantes, UMR 6183 CNRS, Nantes, France
e-mail: rian.seghir@ec-nantes.fr

F. Pierron · L. Fletcher
Engineering and the Environment, University of Southampton, Highfield Campus, Southampton, UK

A. Baldi et al. (eds.), *Residual Stress, Thermomechanics & Infrared Imaging, Hybrid Techniques and Inverse Problems, Volume 7*, Conference Proceedings of the Society for Experimental Mechanics Series, https://doi.org/10.1007/978-3-319-95074-7_19

19.2 Theoretical Framework

The main idea behind the proposed IBSR methodology is to replace a standard *a priori* parametrisation of the material "law" by a piecewise spatial parametrisation of the acceleration and by extension of the stress components through dynamic equilibrium. This is accomplished using the general rules that the field of stress tensor has to satisfy regardless of its constitutive equation, *i.e.* conservation laws, linear and angular momentum, boundary conditions and continuity. Here, the acceleration fields are expanded using a bi-linear element formulation leading to a quadratic form of the piecewise components of the stress tensor field. As a consequence, equilibrium provides six independent equations per element, imposing cross-slopes and curvatures. Continuity provides 18 independent equations, connecting elements state. Applying boundary conditions provide the constants. Finally, an overdetermined linear system of equation is assembled and can be solve in a least-square sense if acceleration fields, material density and boundary conditions are measured/known. In the following, the material density is considered homogeneous and reasonably independent from loading rate and temperature. Acceleration fields are obtained by double differentiation of displacements output from explicit dynamic simulations (note that these are currently available experimentally, see [2]). The boundary conditions are partially known due the test configuration. Here, we focus here on an inertial impact test configuration [2] where the free boundary conditions impose the normal and the tangent components of the stress tensor to be zero at all times at the specimen borders. Unknown boundary conditions, i.e. at impact edge, are eventually recovered by iteratively back identifying the specimen tangent stiffness map and updating local stresses components from strains.

19.3 Simulation

The model consists in a 54 mm $\times$ 34 mm rectangular specimen, meshed with 136 $\times$ 216 4-nodes quadrilateral plan stress elements. The loading is introduced by explicitly simulating the impact between the specimen and a 10 mm long steel projectile at a speed of 45 m/s. The specimen material is isotropic linear elastic and homogeneous with properties as follows: $\rho = 14{,}000$ kg.m^{-3}, $Q_{11} = 580$ GPa, $Q_{12} = 140$ GPa, $Q_{66} = 210$ GPa. The explicit dynamic simulation is performed using LS-Dyna and results are output with a time-step of 200 ns (equivalent to an acquisition rate of 5 MHz). It is worth noting that while the material is homogeneous, stress fields are neither uniform nor constant. Indeed, the dynamic loading produces waves that propagate back and forth in the specimen interfering each other. For the stress reconstruction, a mesh size has to be chosen. A finer mesh will produce better reconstruction. Nevertheless, elements have to be large enough to identify the set of parameters, here 5 per stress component. Hence, in the following a stress mesh size of 5 $\times$ 5 FE elements is chosen.

19.4 Results

Figure 19.1a, b show the error, mean and standard deviation over the field, for the reconstructed stress and tangent stiffness components at a specific time step $t = 16$ µs. Figure 19.1c presents the associated equilibrium gap along the x and y axis, *i.e.* the difference between the acceleration fields multiplied by the density and the divergence of the reconstructed stress tensor fields. Note that in the following, identification windows of 15 $\times$ 15 FE elements (3 $\times$ 3 stress elements) are used to iteratively back identify the tangent stiffness maps. Iterations correspond to successive updates of the boundary conditions starting the assumption that edges are free.

The average value of the equilibrium gaps shows that the reconstructed stress fields are well balance with the acceleration fields at all stress elements. The standard deviation is directly dependent on the parametrisation of the acceleration field. Indeed, only bi-linear variations over a 5 $\times$ 5 FE element can be captured nevertheless its remains negligible compared to the overall signal, about 10^{10} N.m^{-3}. Both results confirm that the reconstructed stress tensor fields are fully statically admissible. While a large standard deviation is observed for the reconstructed stress components at the first iteration (*i.e.* with the assumption that all edges are free), the error rapidly decreases reaching $-1^{\pm 12}$MPa, $0^{\pm 11}$MPa and $0^{\pm 5}$MPa for the σ_{xx}, σ_{yy} and σ_{xy} respectively (*i.e.* less than 2% of the input signal). Similarly, tangent stiffness components converge rapidly and the error reaches $2^{\pm 5}$%, $1^{\pm 6}$%, $2^{\pm 22}$%, $1^{\pm 8}$% for Q_{11}, Q_{22}, Q_{12} and Q_{33} respectivelly. Q_{13} and Q_{23} are found to be equal to $0^{\pm 20}$GPa. The average behaviour is perfectly captured, within 2% of error, as well as the material symmetry. Locally, 5% variations are found on the more active parameters, but 20% error is found on quasi-null or poorly actived parameters. It is important to notice that any a priori knowledge on the material symmetry would significantly decrease the uncertainty.

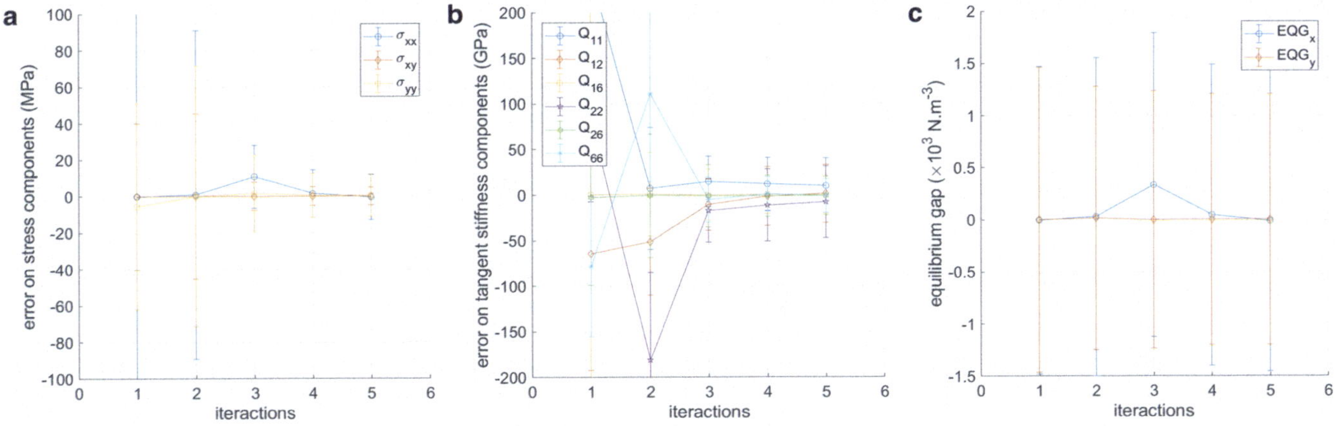

Fig. 19.1 Identification error as a function of iteration number, for a stress mesh of 5x5 FE elements at time t = 16 μs. (**a**) Error on stress, (**b**) on tangent stiffness and (**c**) associated equilibrium gap

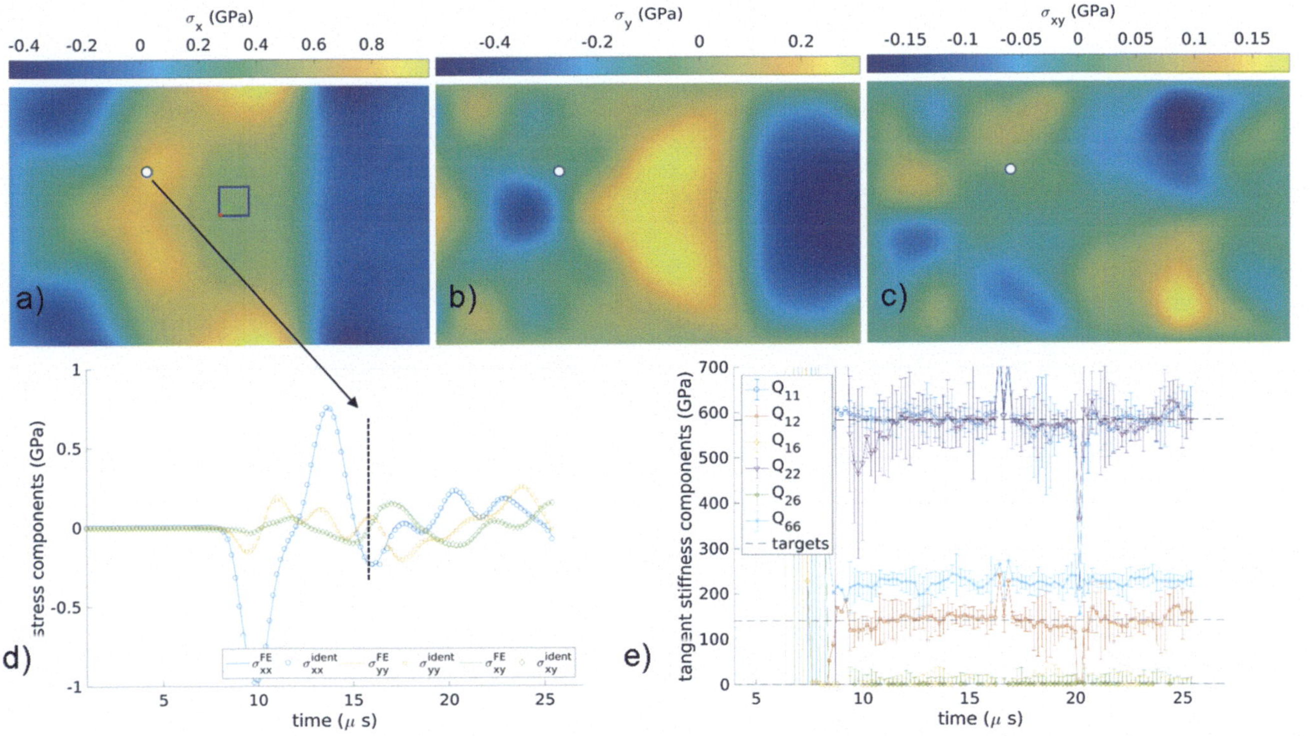

Fig. 19.2 From (**a–c**) identified stress fields at 16 μs. (**d**) FE vs reconstructed stress at a specific point over time, and (**e**) identified material properties over time

Here the six components of the stiffness tensor are treated as fully independent. Another solution would be to enlarge the identification window to find an improved compromise between the desired spatial resolution of the material property fields and their accuracy. Finally, it is worth noting that some of the error found here, is due to the FE calculation itself. Indeed, the FE mesh size has been chosen to provide a good compromise between calculation time and accuracy, nevertheless, by simply checking the FE equilibrium gap, an error of $0.5^{\pm 3}\%$ can already be found. A mesh refining analysis is under progress and is expected to significantly decrease the scattering on the identification.

Figure 19.2a–c present the converged (5th iteration) stress fields (same time step as in Fig. 19.1). Figure 19.2b shows the comparison between FE and reconstructed stresses over time at a specific point (marked with a white dot) and Fig. 19.2e shows the variability over time and space of the identified material properties (mean and standard deviation). The stress fields are highly heterogeneous and Fig. 19.2d clearly shows the ability of the proposed methodology to accurately capture the variation of the stress field in space and time. Figure 19.2e shows that the material properties are reasonably well captured over time. Note that some time steps lead to a higher error due to the decrease in the input signal (*i.e.* low acceleration).

19.5 Conclusion

The preliminary results of this study validate the IBSR methodology. This shows that it is possible to reconstruct the full stress tensor from the acceleration field and an appropriate test configuration. Experimental data, for isotropic and orthotropic material, are currently being processed with the IBSR method and will confirm the use of the method for real experimental data. The IBSR method will be especially useful for accessing strength in an off-axis impact test on a composite specimen as it will allow for the full stress tensor to be reconstructed and rotated into the material co-ordinate system. A local stress-strain response can then be derived at the fracture location allowing for direct identification of the stress-state at failure.

References

1. Pierron, F., Forquin, P.: Ultra-high-speed full-field deformation measurements on concrete spalling specimens and stiffness identification with the virtual fields method. Strain. **48**(5), 388–405 (2012)
2. Pierron, F., Zhu, H., Siviour, C.: Beyond Hopkinson's bar. Phil. Trans. R. Soc. A. **372**(2023), 20130195 (2014)
3. Seghir, R., Pierron, F.: A novel image-based ultrasonic test to map material mechanical properties at high strain-rates. Exp. Mech. **58**(2), 183–206 (2018)

Chapter 20
Infrared Thermography for Material Characterization at Intermediate Strain Rates

Thomas Fourest and Julien Berthe

Abstract The presented activity consist in the measurement of the self-heating field during testing of a metallic material from quasi-static to intermediate strain rates (10^{-2} s^{-1}–100 s^{-1}). It is expected that for these strain-rates the self-heating process will be between the isothermal and adiabatic domains. The investigated material is a dual-phase DP450 steel. The modelling of the self-heating is of great importance for the modelling of the intermediate strain-rates behaviour of metallic material, especially for complex and non-monotonic loading.

Keywords Self-heating · Strain rate effects · Material characterisation · Viscoplasticity · High strain rate tests

20.1 Introduction

It is well known that the behavior of most metallic material is dependent on strain rate. In order to perform numerical simulations of crash incidents, the material behavior must be characterized at intermediate strain rate (1 s^{-1}–100 s^{-1}). If the material flow stress is dependent on the temperature, then the stress flow recorded for an isotherm (low strain-rate) test and an adiabatic (high strain-rate) test will not be the same. In fact, the differences in the stress flow are due to both the temperature and the strain-rate effect. Practically differentiation of the two effects is usually not done and the evolution attributed to the strain rate include the softening due to the temperature rise. This approach is reasonable for monotonic loading for which the material in the structure will behave as it did during the characterisation tests. However Kapoor and Nemat-Nasser [1] showed that for non-monothonic loading in which the specimen is allowed to cool down before reloading , this approach results in an underestimation of the stress flow for the second loading. An approach to separate the contribution of the temperature from the strain-rate one in elasto-viscoplastic models, is to model the self-heating as an internal variable, as a function of the plastic deformation as in Eq. (20.1) in [2].

$$\dot{T} = \omega(\dot{\varepsilon}_p).\chi.\frac{\sigma_{eq}\dot{\varepsilon}_p}{\rho C_p} \tag{20.1}$$

where $\dot{T}$ is the temperature increase rate, χ the Taylor-Quinney parameter which gives the part of plastic work converted to heat, ρ and Cp are the specific mass and the specific heat of the material, σ_{eq} is the equivalent stress and $\omega(\dot{\varepsilon}_p)$ is a strain-rate dependent factor. Some authors as Mason et al. [3] have investigated experimentally the dependency of $\omega(\dot{\varepsilon}_p)$ and χ on the strain-rate. Most experimental research on this field has been performed on Split Hopkinson Pressure Bar, so not for intermediate strain rates. However, this approach may enable some separation between the two effects during characterization tests but it does not deal with the previously stated limitations if no thermomechanical computation is used.

Therefore, it is important to increase the knowledge on self-heating in medium-strain rate tests to better know the domain of validity of the previously describe type of models and to know if thermomecanical calculations are needed to better predict crash situations. This study uses infrared thermography as a mean to measure the local specimen heating. Lagrangian thermography will be used to follow the deformation and heating of material points during the characterization tests. In this approach local heating and deformation can both be used for the identification of the material parameters.

T. Fourest (✉) · J. Berthe
Department of Materials and Structures, French Aerospace Research Center (ONERA), Lille, France
e-mail: thomas.fourest@onera.fr

A. Baldi et al. (eds.), *Residual Stress, Thermomechanics & Infrared Imaging, Hybrid Techniques and Inverse Problems, Volume 7*, Conference Proceedings of the Society for Experimental Mechanics Series, https://doi.org/10.1007/978-3-319-95074-7_20

20.2 Characterisation Tests on DP450 Steel

Characterisation tests have been performed using a Schenk hydraulic jack capable of reaching a 10 m/s loading speed with a capacity of approximately 20 kN. Several characterisation tests have been performed at loading speed from 5 mm/min to the limit of exploitation of the hydraulic jack tests which is around 1 m/s for this type of specimen. The strain on the surface of the specimen is measured using Digital Image Correlation (DIC) for images recorded with a Photron SA-X camera that can reach 12,000 frames per second with a resolution of 1 Mpixel. The temperature fields are recorded using a CEDIP infrared camera on the opposite side of the specimens. The frequency of the infrared camera is lower than the one of the visible camera, however for the higher loading speed (1 m/s) it was still possible to record more than 20 frames during the loading with a resolution 80 × 110 pixels. Figure 20.1 shows the devices used in these tests.

Figure 20.2 compares the load-displacement curves for the three loading speeds tested. It is observed, from this information alone, that the material seems to have an elasto-viscoplastic behavior: the rigidity of the specimen does not evolve but a significant change appears in the non-linear part of the curve in function of the loading speed. The strain fields will be computed using DIC.

Fig. 20.1 Devices used for the characterization tests: hydrolic jack, infrared camera and high-speed camera

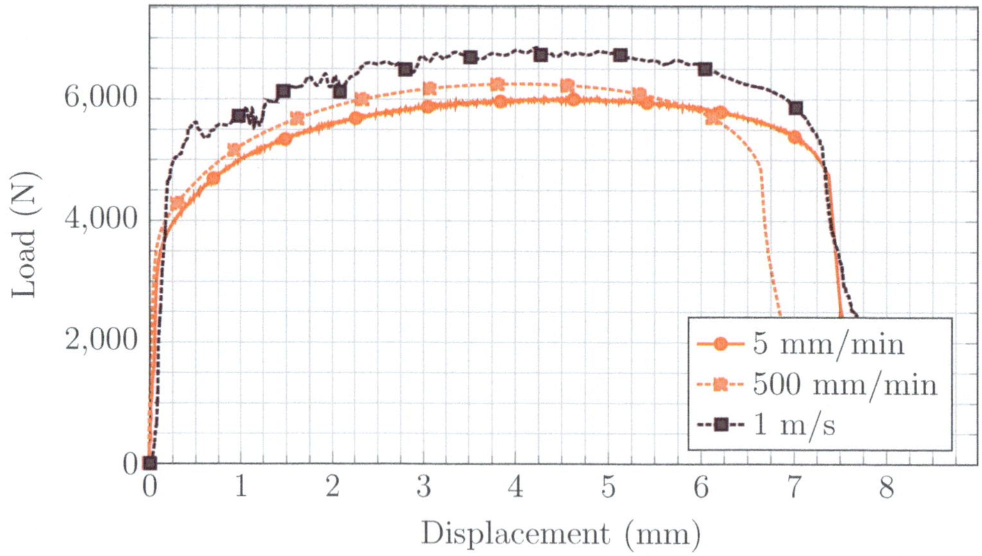

Fig. 20.2 Load-Displacement curves for the characterisation specimen at loading speed 5 mm/min, 500 mm/min, and 1 m/s

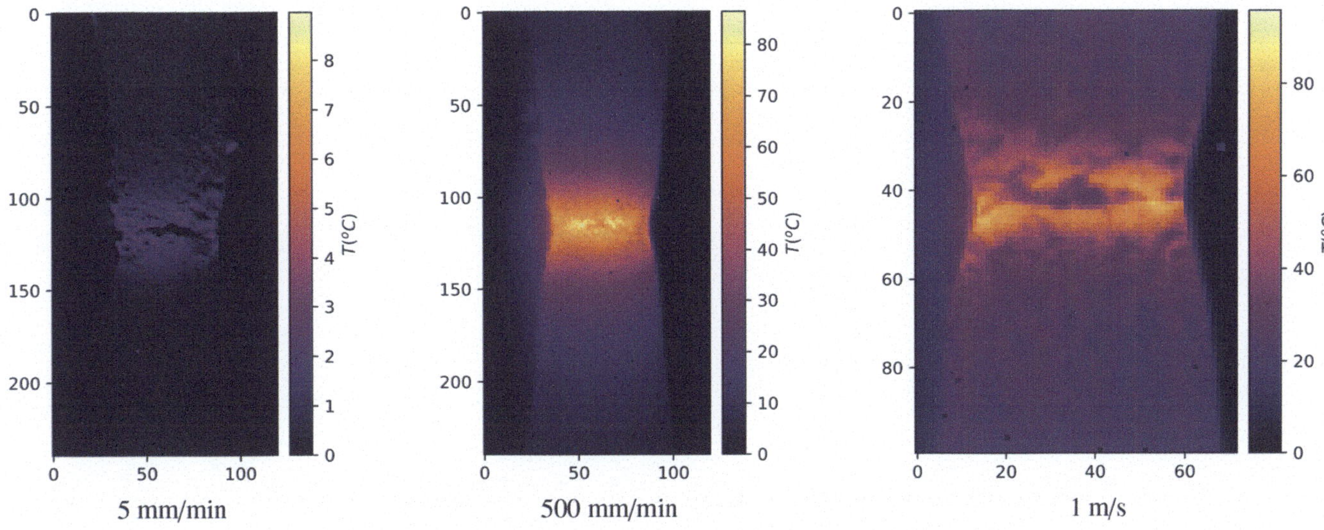

Fig. 20.3 Typical temperature increase fields measured before the failure of the specimens for each loading speed

20.3 Infrared Measurement

Figure 20.3 shows the self-heating fields recorded just prior to the failure of the characterisation specimens for the three loading speeds. From these measurements, it is seen that the temperature increase is only of a few Celsius degree for a 5 mm/min loading speed, of approximately 80 °C at 500 mm/min and of approximately 100 °C at 1 m/s. From these test it is concluded that the final temperature is very dependent on the strain-rate. It can also be seen that the self-heating is very localised even on a unidirectional specimen. Future comparison with the strain fields will give more information on the nature of the heating process (isothermal, adiabatic, . . .).

20.4 Conclusion

In this study full kinematic and thermal field measurement have been performed during characterisation tests of a DP450 steel for loading speed ranging from 5 mm/min to 1 m/s (which results in strain-rates from $10^{-}2\ s^{-1}$ to $100\ s^{-1}$). It has been observed that self-heating is very dependant on the strain-rate. It goes from a few degrees for quasi-static speed to 100 °C for the higher strain-rate.

References

1. Kapoor, R., Nemat-Nasser, S.: Determination of temperature rise during high strain rate deformation. Mech. Mater. **27**, 1 (1998)
2. Roth, C.C., Mohr, D.: Effect of strain rate on ductile fracture initiation in advanced high strength steel sheets: experiments and modeling. Int. J. Plast. **56**, 19 (2014)
3. Mason, J., Rosakis, A., Ravichandran, G.: On the strain and strain rate dependence of the fraction of plastic work converted to heat: an experimental study using high speed infrared detectors and the Kolsky bar. Mech. Mater. **17**, 135 (1994)

Chapter 21
Evaluation of Fatigue Damage in Short Carbon Fiber Reinforced Plastics Based on Thermoelastic Stress and Phase Analysis

Takahide Sakagami, Daiki Shiozawa, Yu Nakamura, Shinichi Nonaka, and Kenichi Hamada

Abstract Short carbon fiber composite materials are receiving a lot of attentions because of their excellent moldability and productivity, however they show complicated behaviors in fatigue fracture due to the random fibers orientation. In this study, thermoelastic stress analysis (TSA) using an infrared thermography was applied to the evaluation of fatigue damage in short carbon fiber composites. Second harmonic component of thermoelastic temperature change that is obtained by lock-in processing based on double-frequency against loading frequency was conducted to identify the turbulence in thermoelastic waveform due to fatigue damage evolution. It was found that the portions showing high second harmonic component values coincided with the portions where delamination damages were detected.

Keywords Nondestructive evaluation · Thermoelastic stress analysis · Second harmonic analysis · Infrared camera · Short carbon fiber reinforced plastics

21.1 Introduction

Carbon fiber reinforced plastic (CFRP) has been widely employed for the primary structural members of transportation vehicles such as automobile, aircraft or spacecraft, utilizing its excellent specific strength and specific rigidity in contrast with the metal. Short carbon fiber composite materials have been receiving a lot of attentions because of their excellent moldability and productivity. However they show complicated behaviors in fatigue fracture due to their random fibers orientation. Therefore effective nondestructive evaluation techniques are required for detecting and measuring various types of damages, such as fiber breakage, matrix cracking and delamination, during fatigue fracture. Nondestructive evaluation (NDE) techniques based on thermoelastic stress analysis (TSA) [1, 2] using infrared thermography has been effectively employed for the detection of delamination defects in fiber reinforced plastics. Thermoelasticity for orthotropic materials have been studied by many researchers and the TSA technique has been employed as a powerful tool for evaluating impact or fatigue damages in composite materials and structures. Emery and Dulieu-Barton [3] applied the TSA technique to fatigue damage evaluation in laminated glass fiber epoxy materials, in which TSA was employed as powerful tool for analyzing complicated fatigue damages such as fiber breakage, matrix cracking and delamination damage in composites. Fruehmann and Dulieu-Barton [4] applied the TSA technique for the assessment of fatigue damage evolution in woven composite materials. Uenoya and Fujii [5] developed thermoelastic damage analysis (TDA) for the early damage detection in plain-woven CFRP. In this technique, a differential thermoelastic temperature distribution image was generated by subtracting an image of thermoelastic temperature change obtained at certain loading cycle from a reference initial image of thermoelastic temperature change. Fatigue damage evolution causes local stress change, and this is emphasized in the differential thermoelastic temperature distribution image. Therefore the fatigue damage in CFRP sample can be detected at the early stage in the fatigue life.

In the previous paper, the present authors [6, 7] investigated the relationship between the fiber orientation angles and the thermoelastic temperature change and its phase delay from the applied loading signal for the short carbon fiber reinforced plastics during fatigue test. CFRP is composed with carbon fibers having negative thermoelastic constant [8] and matrix resin having positive thermoelastic constant. Thus CFRP exhibits temperature change superposing in-phase thermoelastic temperature change in CFRP and out-of-phase thermoelastic temperature change in matrix resin. The characteristics of the

T. Sakagami (✉) · D. Shiozawa · Y. Nakamura
Department of Mechanical Engineering, Kobe University, Nada, Kobe, Japan
e-mail: sakagami@mech.kobe-u.ac.jp

S. Nonaka · K. Hamada
DIC Corporation, Chuo-ku, Tokyo, Japan

A. Baldi et al. (eds.), *Residual Stress, Thermomechanics & Infrared Imaging, Hybrid Techniques and Inverse Problems, Volume 7*, Conference Proceedings of the Society for Experimental Mechanics Series, https://doi.org/10.1007/978-3-319-95074-7_21

superposed thermoelastic temperature change are influenced by the load sharing condition between carbon fibers and matrix resin, locally changed due to the fatigue damage evolution or fiber orientation angles. The present authors conducted fatigue damage evaluations based on the conventional TDA procedure as well as the newly developed phase-delay based damage characterizing technique "thermoelastic phase damage analysis (TPDA)".

In this study, the authors focused on the second harmonic component of the thermoelastic temperature change to detect fatigue damage evolution in short carbon fiber composites. Paynter and Dutton [9] investigated the nonlinear effects in thermoelasticity caused by the presence of damage, and applied the TSA technique to wind turbine blade composite structure with successful results on damage evaluation using second harmonic signal correlation. It is expected that the short carbon fiber composites show characteristic thermoelastic temperature change that reflects the change in load sharing condition between carbon fibers and matrix resin due to fatigue damage evolution and fiber orientation angles. The second harmonic component analysis is employed to extract the change in thermoelastic temperature change in fatigue test for the short carbon fiber composites. The relationship between the second harmonic component obtained by double frequency lock-in processing and fatigue damages are investigated for several stages of loading cycles.

21.2 TSA and Second Harmonic Component of Thermoelastic Temperature Change Using Infrared Thermography

Dynamic stress change cause a very small temperature change under adiabatic conditions in a solid. This phenomenon is known as the thermoelastic effect and is described by Lord Kelvin's equation, which relates the temperature change (ΔT_{E}) to the sum of the changes in the principal stresses ($\Delta\sigma$) under cyclic variable loading as follows.

$$\Delta T_{\mathrm{E}} = -\frac{\alpha}{\rho C_p} T\Delta\sigma = -kT\Delta\sigma \tag{21.1}$$

Here, α is the coefficient of thermal expansion, ρ is the mass density, C_p is the specific heat at constant pressure and T is the absolute temperature. The coefficient k is called thermoelastic constant. The sum of the changes in the principal stresses ($\Delta\sigma$) is obtained by measuring the temperature change (ΔT_{E}) using infrared thermography.

As thermoelastic temperature changes are very small and sometimes hidden by the thermal noise of the infrared camera, lock-in infrared thermography using reference signals synchronized with the stress changes is commonly employed to improve the accuracy of stress measurements. The TSA technique uses a lock-in algorithm with a reference-loading signal extracted from the load cell or strain gauge to improve the signal–noise ratio.

As described in the foregoing paragraph, fatigue damage evolution influences not only on the amplitude of thermoelastic temperature change but also on the waveform of thermoelastic temperature change. This may generate harmonic components in the waveform even under the sinusoidal waveform loading. In this experiment the second harmonic component of thermoelastic temperature change was extracted from surface temperature change measured by infrared thermography based on the Fourier analyses.

21.3 Experimental Study

21.3.1 Specimen and Experimental Setup

Configurations of the CFRP specimen employed in this study are shown in Fig. 21.1. The specimens were cut from laminated short fiber CFRP sheet with vinyl ester resin and 25.4 mm long carbon fiber bundles. Each bundle was composed of 12,000 short carbon fibers. The mass content (wt %) of resin and fiber was 67 and 33, respectively. The specimen has circular notches with a radius of 2 mm.

Cyclic-axis sinusoidal waveform loading with a frequency f of 7 Hz and a stress ratio $R = 0.1$ was applied to the specimen by an electrohydraulic fatigue testing machine. Microscopic visible images on the specimen surface and side surface were taken by optical microscope. The temperature change on the specimen surface was measured by infrared thermography with an MCT array detector (FLIR Systems Inc., SC7500). The specifications and setting of this infrared camera are listed in Table 21.1. The thermoelastic temperature change ΔT_{E}, the phase difference $\Delta\theta_{\mathrm{E}}$ and the second harmonic component were obtained from experimental data as shown in the forgoing paragraph.

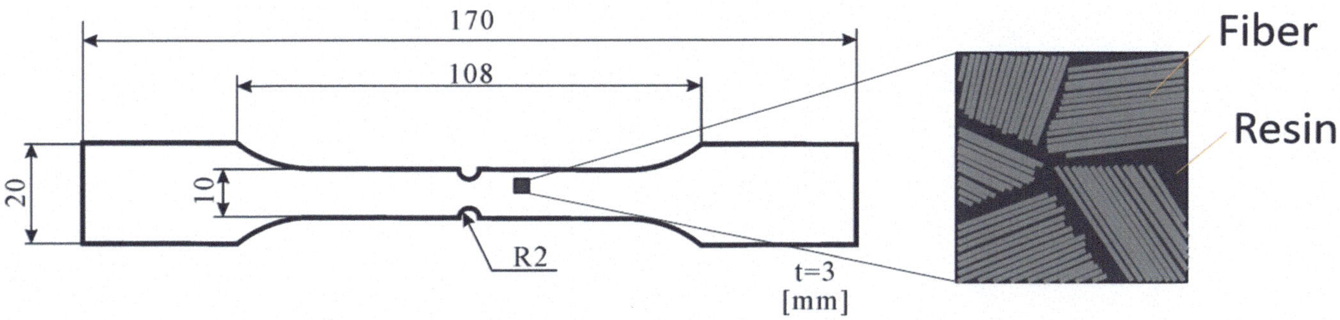

Fig. 21.1 Configurations of employed short carbon fiber reinforced plastic specimen

Table 21.1 Specifications and setting of employed infrared camera

Infrared detector	MCT
Detectable wavelength	7.7–9.3 μm
Number of detectors	320 × 256
Temperature resolution (NETD)	25 mK
Framing rate	373 Hz
Time of data acquisition	10 s

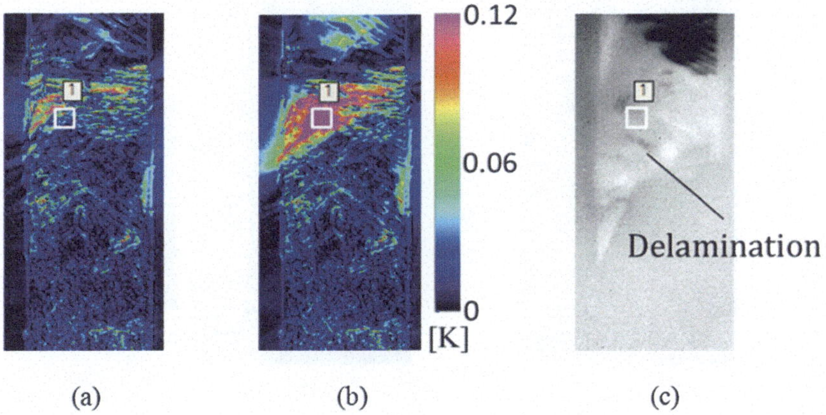

Fig. 21.2 Distribution of second harmonic component and infrared image after fracture. (σ_{max} = 100 MPa). (**a**) 38,000 cycles, (**b**) 40,000 cycles, (**c**) Fracture: N_f = 40,168 cycles

21.3.2 Experimental Results and Discussion

Fatigue test was conducted for the short fiber CFRP specimen with circular notches under the applied maximum stress σ_{max} = 100 MPa. The distributions of second harmonic component values obtained at 38,000 and 40,000 cycles are shown in Fig. 21.2, together with the infrared image of the specimen measured after fracture. It is found in the figures that high values of second harmonic component in thermoelastic temperature change are detected in the region where the delamination defect was detected. Figure 21.3 shows waveforms of the thermoelastic temperature change obtained in the region where the higher values of the second harmonic component (indicated by white square mark in Fig. 21.2a, b), with the waveform of loading signal. It is found in the figure that the sinusoidal thermoelastic temperature change is observed around 38,000 cycles, and the waveform is out-of-phase against the loading signal. In contrast, the waveform of thermoelastic temperature change observed around 40,000 cycles is distorted from sinusoidal waveform when the maximum tensile load is applied. It seems that this kind of nonlinear behavior in thermoelastic temperature change is caused by the evolution of fatigue delamination damage.

Another experimental result indicating the effectiveness of second harmonic component measurement for damage identification is shown below. Fatigue test was conducted for another short fiber CFRP specimen under the applied maximum stress σ_{max} = 180 MPa. Figure 21.4 shows the distributions of second harmonic component values obtained at 200 and 4000 cycles, before the specimen failure occurred at 44,221 cycles. It is found from the figures that localized high second harmonic value regions are detected. Two evaluation regions were set in the high value area #1 as well as in the low value

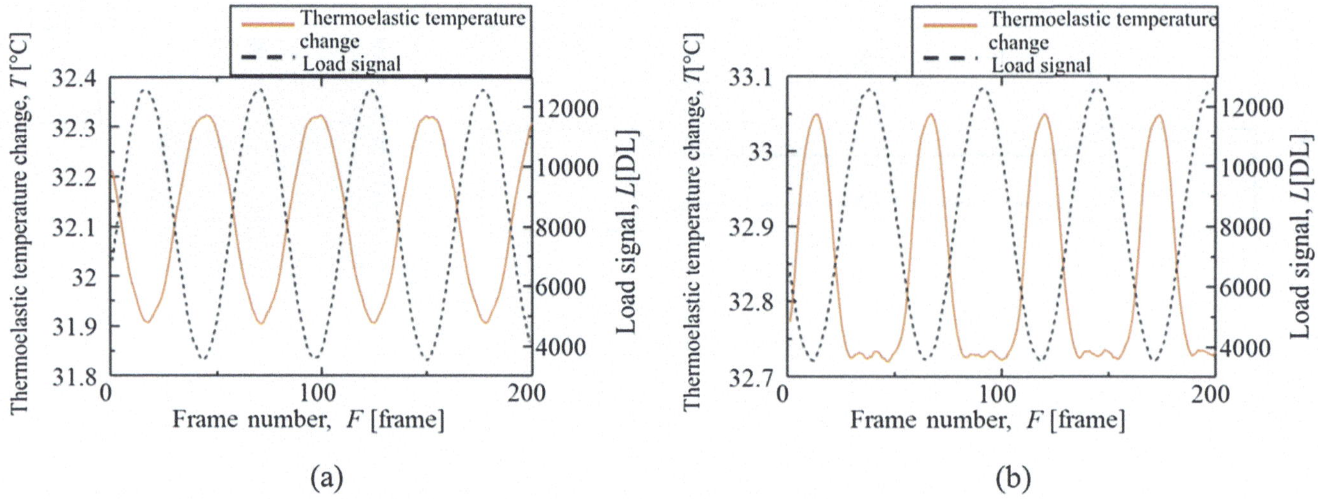

Fig. 21.3 Observed waveforms of thermoelastic temperature change and loading signal. ($\sigma_{max} = 100$ MPa), (**a**) 38,000 cycles, (**b**) 40,000 cycles

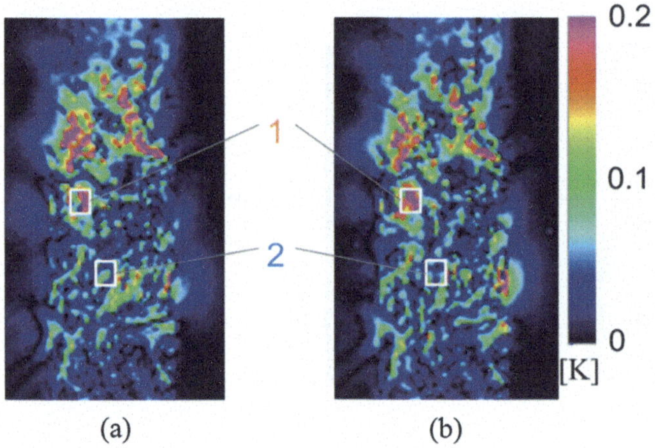

Fig. 21.4 Distributions of second harmonic component. ($\sigma_{max} = 180$ MPa). (**a**) 200 cycles, (**b**) 4000 cycles

area #2 as indicated in Fig. 21.4. The waveforms of thermoelastic temperature change obtained in area #1 and area #2 are shown in Fig. 21.5. It can be seen from the figures that thermoelastic temperature changes being in-phase against loading signal are observed, in contrast with the out-of-phase temperature changes shown in Fig. 21.3. As described in the previous papers [6, 7], this in-phase thermoelastic temperature change means that the thermoelastic effect of carbon fiber bundles is predominant. In area #1 where the higher second harmonic values are observed, the difference in thermoelastic temperature change becomes larger, and its waveform at the maximum loading is distorted from sinusoidal waveform compared with that at the minimum loading. Further it is found from Fig. 21.5b that the waveform at the maximum loading shows more significant distortion at 4000 cycles ($N/N_f = 0.1$) and thermoelastic temperature change shows plateauing above certain loading level.

It is found from the experimental results that nonlinear behavior is observed in the waveform of thermoelastic temperature change at the timing of maximum tensile loading whether the thermoelastic temperature change is out-of-phase or in-phase against the loading signal. This nonlinear behavior in thermoelasticity generates the second harmonic component that can be detected by the double-frequency lock-in data processing. It is well known as energy dissipation that second harmonic component in thermoelastic temperature change appears due to local plasticity in metals. However it is unlikely to understand that plastic deformation occurs in carbon fibers or matrix thermohardening resin themselves. Further the nonlinear behavior is observed only at the maximum tensile loading, not observed in unloading process. Therefore it can be concluded that the nonlinear behavior in thermoelasticity is caused by the evolution of delamination fracture between carbon fibers and matrix resin.

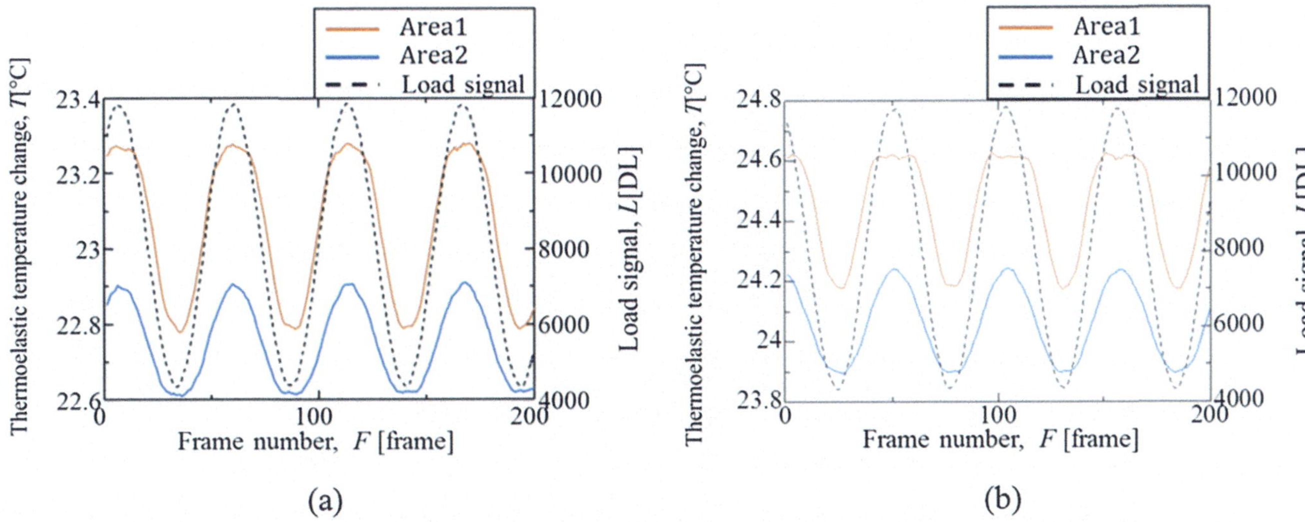

Fig. 21.5 Observed waveforms of thermoelastic temperature change and loading signal. (σ_{max} = 180 MPa). (**a**) 200 cycles, (**b**) 4000 cycles

21.4 Conclusions

In this study, appearance of the second harmonic component in thermoelastic temperature change was investigated for short carbon fiber composites in fatigue test. Localized high second harmonic value regions were detected in fatigue test. These regions were coincided with the portion where the delamination damage was detected. It was also found that the waveforms of thermoelastic temperature change observed in these areas were distorted from sinusoidal waveform at the timing of maximum loading was applied. It can be concluded that the nonlinear behavior in thermoelasticity was caused by the evolution of delamination fracture between carbon fibers and matrix resin.

References

1. Greene, R.J., Patterson, E.A., Rowlands, R.E.: Thermoelastic stress analysis. In: Sharpe Jr., W.N. (ed.) Springer Handbook of Experimental Solid Mechanics, pp. 743–767., ISBN 978-0-387-26883-5. Springer Science + Business Media, LLC, New York (2008)
2. Dulieu-Barton, J.M.: Introduction to thermoelastic stress analysis. Strain. **35**, 35–39 (1999)
3. Emery, T.R., Dulieu-Barton, J.M.: Thermoelastic stress analysis of damage mechanisms in composite materials. Compos. Part A. **41**, 1729–1742 (2010)
4. Fruehmann, R.K., Dulieu-Barton, J.M., Quinn, S.: Assessment of fatigue damage evolution in woven composite materials using infra-red techniques. Compos. Sci. Technol. **70**, 937–946 (2010)
5. Uenoya, T., Fujii, T.: Damage characterization of woven fabric composite materials by thermoelastic analysis. J. Soc. Mater. Sci. Jpn. **49**, 941–947 (2000)
6. Sakagami, T., Shiozawa, D., Nakamura, Y., Nonaka, S., Hamada, K.: Fatigue damage evaluation of short fiber CFRP based on phase information of thermoelastic temperature change. Proc. SPIE. **10214**, 102140M-1–102140M-6 (2017). https://doi.org/10.1117/ 12.2262972
7. Shiozawa, D., Sakagami, T., Nakamura, Y., Nonaka, S., Hamada, K.: Fatigue damage evaluation of short carbon fiber reinforced plastics based on phase information of thermoelastic temperature change. Sensors. **17**(12), 2824 (2017). https://doi.org/10.3390/s17122824
8. Sugimoto, S., Ishikawa, T.: Examination of quantitative infrared stress measurement of CFRP laminates and its application to non-destructive evaluation. Tech. Rep. National Aerosp. Lab. **1396**, 1–23 (1999)
9. Paynter, R.J.H., Dutton, A.G.: The use of a second harmonic correlation to detect damage in composite structures using thermoelastic stress measurements. Strain. **39**, 73–78 (2003)

Chapter 22
Thermographic Stress Field Investigation of a Multiply-Loaded Disk

Mohammad Yousefi, Xavier Balandraud, and Wael A. Samad

Abstract Hybrid thermoelastic stress analysis (Hybrid-TSA) is an experimental thermographic method that has been successfully utilized for the stress analysis of numerous structures with various geometries, discontinuities and loading situations. Previous work has shown the capacity of such approach to separate stresses on diametrically loaded disks with known loading conditions. The objective of the present work is to investigate the capacity of such hybrid experimental-analytical approach in the stress analysis of two-dimensional granular materials. Previously, thermography has been successful at determining the hydrostatic stress network in cohesionless bidisperse granular systems (composed of cylinders placed in parallel) under confined compression. However, the analysis remained tricky because of the large number of cylinders considered, leading to a reduced number of thermal data per cylinder. The method which is proposed here relies heavily on analytical expressions of stress, which arise from mechanical compatibility and equilibrium conditions. It enables us to reconstruct a stress field on granular materials with low special resolution (few TSA points available per cylinder), and thus to analyze more precisely the mechanical state of tested granular systems.

Keywords Thermography · Granular materials · Hybrid-TSA · Multiple loading

22.1 Introduction

Granular materials are among the materials which do not fall within the traditional definitions of matter. They are composed of "large conglomerations of discrete macroscopic particles" and do not behave like any other forms of matter [1]. Thus, due to their granular nature, regular macroscopic analysis may not be applied and the behavior of the material is best determined using a grain scale analysis [2]. The proper analysis of such material is of great importance due to their increasing usage in many industries such as the geotechnical, chemical, pharmaceutical, agronomical and civil engineering [3]. The behavior of granular materials is of great importance in fields such as chemical and agronomical industries since many of the materials used are prepared from powders or grains. However, the importance of the behavior granular materials while under stress is of greater importance in fields such as civil and geotechnical engineering. This is due to the fact that large stresses will be placed on the granular materials (e.g. soil), for which the effects of which such as internal stresses and possible deformation need to be known during design and construction [4]. Extensive research has been done to study and quantify the effects of stresses on various granular materials, specifically the stress propagation within the material. The stress distribution within the material is identified by the stress paths (or force chains) which are generated within the material when it is put under compression. These stress paths are complex and are dependent on the placement of the grains relative to one another and more often than not, are random due to the random placement of the grains within the material [5]. The knowledge of stresses within the material is important as the stress concentrations that may arise can lead to high deformation of the grains (and particle collapse in some cases), and cause large deformation of the material as a whole. The knowledge of the stress paths also allows us to identify the force propagation within the material and to the material boundary, this knowledge allows for the calculation of the Janssen Constant, which is important when determining pressure on the walls of grain filled silos and an important factor to consider when designing such vessels [5]. Various methods have been employed to study the stresses within granular materials, the two broad categories of these studies are experimental methods and numerical methods. Experimental methods comprise Magnetic Resonance imagery (MRI), X-ray tomography, radar-based

M. Yousefi · W. A. Samad (✉)
Rochester Institute of Technology, Dubai, UAE
e-mail: wascad@rit.edu

X. Balandraud
Université Clermont Auvergne, CNRS, SIGMA Clermont, Institut Pascal, Clermont-Ferrand, France

A. Baldi et al. (eds.), *Residual Stress, Thermomechanics & Infrared Imaging, Hybrid Techniques and Inverse Problems, Volume 7*, Conference Proceedings of the Society for Experimental Mechanics Series, https://doi.org/10.1007/978-3-319-95074-7_22

sensing, positron emission particle tracking, photoelasticimetry and infrared (IR) thermography [6]. Among these methods, IR thermography has proven to be a new choice for full field study of the stresses of granular materials. Under cyclic loading conditions and appropriate thermal data processing, it can give a good image of the full stress field within the material [3]. Many approaches have been taken to numerically identify the stresses within granular materials. These approaches include multi-line simulations of the material using multi plane patterns [7] and discrete element analysis, where the behavior within the material as well as its external behavior on other objects is investigated [4, 8, 9]. These numerical methods are used to generate results which are then verified through experimental means. Acquiring such results is difficult at smaller grain sizes due to the limitations of sensing equipment, thus to acquire an accurate grain and field level information, numerical analysis needs to be coupled with experimental measurements.

22.2 Present Work

Presented here is a hybrid approach for modeling granular systems relying on exact stress solutions of contact problems, coupled with boundary and loading conditions from experimental data. The analytical component utilizes the *Flamant* solution from *theory of elasticity* to describe stress fields of contact regions on flat surfaces, while the experimental data utilizes TSA to locate and evaluate the relative magnitudes of the contact stresses. The overall reconstruction of the stress fields is done via the superposition of the various contact solutions applied to any particular disk. Shown here is a case adopted with permission from [10], where three cylinders (disks in 2D) are cyclically loaded at a frequency of 3 Hz and at room temperature with the TSA image at the center of Fig. 22.1 below. The three disks are made of polyoxymethylene and are 61 mm in diameter. The analytical solution here can be thought of as a superposition of multiple Flamant solutions of contact stresses, Eqs. (22.1, 22.2 and 22.3), along with a radial stress field, Eq. (22.4), where x represents the horizontal coordinate and the y represents the vertical one.

$$\sigma_{x\,Flamant} = \frac{-2Px^2y}{\pi\left(x^2+y^2\right)^2} \tag{22.1}$$

$$\sigma_{y\,Flamant} = \frac{-2Py^3}{\pi\left(x^2+y^2\right)^2} \tag{22.2}$$

$$\tau_{xy\,Flamant} = \frac{-2Pxy^2}{\pi\left(x^2+y^2\right)^2} \tag{22.3}$$

$$\sigma_{x\,radial} = \sigma_{y\,radial} = \frac{2P}{\pi D} \tag{22.4}$$

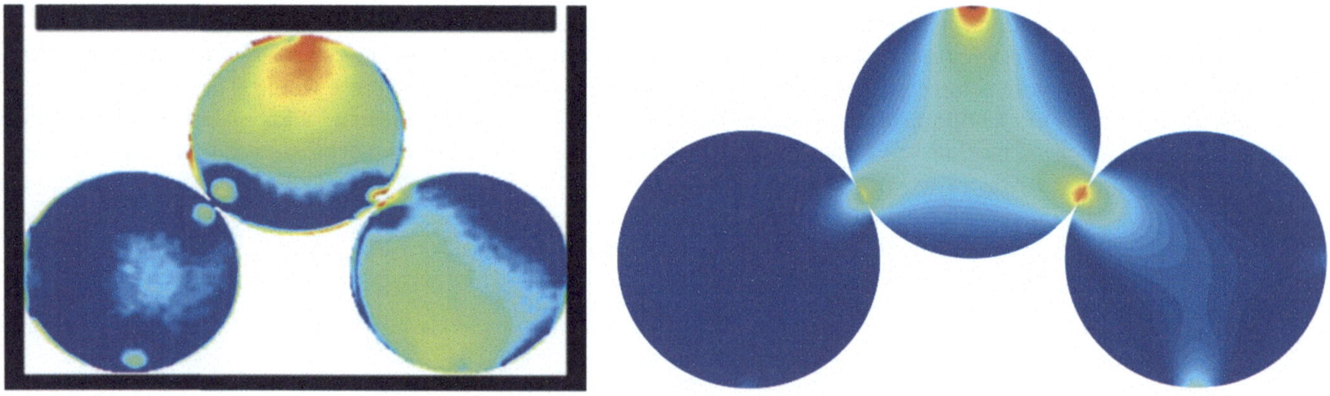

TSA image (S) adopted from [10] *Reconstructed σ_y using proposed hybrid approach*

Fig. 22.1 Reconstructed σ_y using hybrid-TSA for the loading scenario in [10]

The number and positioning of the multiple *Flamant* solution is dependent on the granular material under investigation, and can be determined from the TSA image using imaging and recognition techniques. While this process of contact location identification, and relative S magnitude determination can be automated, here it is performed manually at this stage of research. Figure 22.1 above represents the reconstructed σ_y on all three disks using the hybrid approach presented here. This information is coupled with basic force equilibrium such that the determined relative magnitudes obey static equilibrium. While σ_y is the one shown here in Fig. 22.1, the same can be done for any individual stress component in rectangular $(\sigma_x, \sigma_y, \tau_{xy})$ or polar $(\sigma_r, \sigma_\theta, \tau_{r\theta})$ coordinates, and ultimately a visual of the stress paths (force chains) can be predicted.

22.3 Conclusion

The advantages in the present technique is the ability to reconstruct stress fields (and ultimately stress paths) in granular materials from low resolution TSA images, with the emphasis being on the geometric location of the contacts as well as their relative magnitudes. With the behavior being linear-elastic, exact solutions derived and superposed from *theory of elasticity* are used to draw the stress maps for all three components of stress at a resolution independent from that of the original TSA image. More work is currently under way to better determine the exact stress magnitudes (as opposed to relative values) as well the automation of the contact force location identification; necessary for more complicated granular structures with large number of cylinders.

References

1. Jaeger, H.M., Nagel, S.R., Behringer, R.P.: The physics of granular materials. Phys. Today. **49**(4), 32–38 (1996)
2. Goldenberg, C., Goldhirsch, I.: Small and large scale granular statics. Granul. Matter. **6**(2–3), 87–96 (2004)
3. Chaiamarit, C., Balandraud, X., Preechawuttipong, I., Grediac, M.: Stress network analysis of 2D non-cohesive polydisperse granular materials using infrared thermography. Exp. Mech. **55**, 761–769 (2015)
4. Zhao, H., An, X., Wu, Y., Qian, Q.: DEM modeling on stress profile and behavior in granual matter. Powder Technol. **323**, 149–154 (2018)
5. Puglisi, A.: Janssen effect and the distribution of internal stresses. 14.11.2001. [Online]. Available: http://denali.phys.uniroma1.it/~puglisi/thesis/node5.html. [Accessed 5 Feb 2018]
6. Jongchansitto, P., Balandraud, X., Grediac, M., Beitone, C., Preechawuttipong, I.: Using infrared thermography to study hydrostatic stress networks in granular materials. Soft Matter. **10**(43), 8603–8607 (2014)
7. Sadrnejad, S.A., Saedi Daryan, A., Ziaei, M.: A constructive model for multi-line simulation of granular material behaviour using multi-plane pattern. J. Comput. Sci. **5**(11), 822–830 (2009)
8. Xiao, Y., Zhang, Z., Chen, L., Zheng, K.: Modeling stress path dependency of cyclic plastric strain accumulation of unbound granular materials under moving wheel loads. Mater. Des. **137**(1), 9–21 (2018)
9. Yan, G., Yu, H.-S., McDowell, G.: Simulation of granular material behaviour using DEM. Procedia Earth Planetary Sci. **1**(1), 598–605 (2009)
10. Chaiamarit, C., Balandraud, X., Grédiac, M., Preechawuttipong, I.: Experimental evidence of stress network in a 2D granular material under confined compression by infrared thermography. In: Photomechanics, Montpellier (2013)

Chapter 23
Fatigue Limit Estimation for Single Bead-On-Plate Weld Based on Dissipated Energy Measurement

Daiki Shiozawa, Yuto Ogino, Takaya Washio, Takahide Sakagami, Hideki Ueda, and Taizo Maikino

Abstract The fatigue limit estimation method based on the dissipated energy measurement was applied to the single bead-on-plate weld. The stair-case-like stress level test was conducted to the bead-on-plate specimen. The local concentration of dissipated energy was observed in the weld toe. The change in dissipated energy at this local high dissipated energy point showed sharp increase from the certain stress amplitude, and this stress amplitude coincided with the fatigue limit for the bead-on-plate specimen obtained from *S-N* curves.

Keywords Dissipated energy · Fatigue · Infrared thermography · Fatigue limit · Weld

23.1 Introduction

Mechanical properties for material strength such as fatigue limit are important parameters for structural design. Conventional laboratory fatigue tests based on 10 million stress cycles takes more time and cost. Therefore, fatigue limit estimation based on the dissipated energy measurement using infrared thermography has been getting an increasing attention in various industrial fields [1]. The applicability of this technique for some steel and metal was reported [2–6]. A lot of industrial structures are made by welding. It is important to evaluate the fatigue strength of welding, and it is useful to apply this technique to the fatigue strength evaluation of welding part. In this study, the dissipated energy at the single bead-on-plate specimen was measured during fatigue test, and the fatigue limit estimation method was applied to the fatigue strength evaluation of the single based-on-welding.

23.2 Fatigue Limit Estimation Based on Dissipated Energy

Temperature rise is observed under compressive stress, and temperature fall is observed under tensile stress, as shown in Fig. 23.1. This phenomenon is called as thermoelastic effect. Thermoelastic temperature change ΔT_E is formulated by thermoelastic coefficient k, absolute temperature T, and sum of principal stresses $\Delta\sigma$, as shown in the following equation.

$$\Delta T_E = -kT\Delta\sigma \tag{23.1}$$

The thermo-elastic temperature change ΔT_E is a reversible phenomenon. In the actual case, temperature rise due to irreversible energy dissipation ΔT_D occurs at the maximum tensile stress and at the maximum compressive stress. Thus, the measured temperature change $T(t)$ on the surface includes ΔT_E and ΔT_D. Therefore, temperature change due to dissipated energy ΔT_D can be obtained as the component having double frequency of the load signal using basically Fourier analysis.

In the fatigue limit estimation based on dissipated energy, the dissipated energy is measured for various levels of stress amplitude during short time cyclic loading (staircase-like stress level test). It is known that the change of dissipated energy shows sharp increase from a certain stress level, as shown in Fig. 23.2. The estimation scheme is summarized as follows [7];

D. Shiozawa (✉) · Y. Ogino · T. Washio · T. Sakagami
Department of Mechanical Engineering, Kobe University, Nada, Kobe, Japan
e-mail: shiozawa@mech.kobe-u.ac.jp

H. Ueda · T. Maikino
Nippon Steel & Sumitomo Metal Corporation, Fuso-cho, Amagasaki, Japan

A. Baldi et al. (eds.), *Residual Stress, Thermomechanics & Infrared Imaging, Hybrid Techniques and Inverse Problems, Volume 7*, Conference Proceedings of the Society for Experimental Mechanics Series, https://doi.org/10.1007/978-3-319-95074-7_23

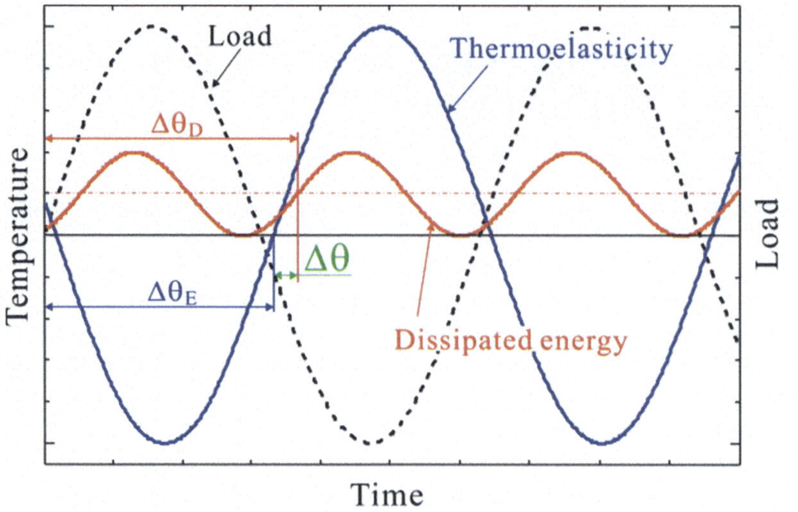

Fig. 23.1 Schematic illustration of temperature change due to the thermoelasticity and energy dissipation

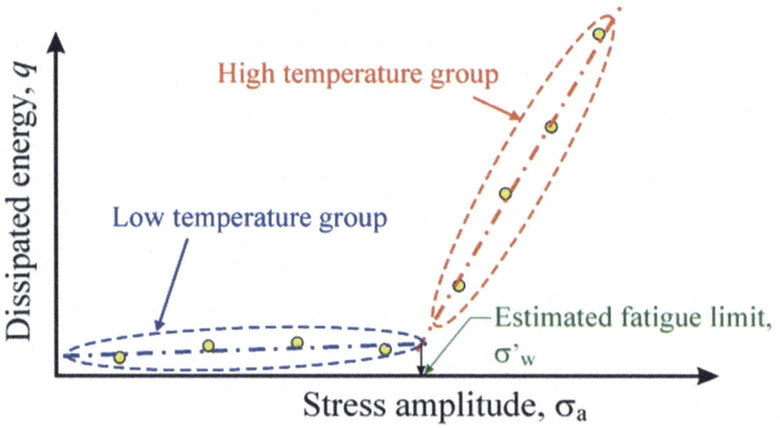

Fig. 23.2 Change of dissipated energy and the fatigue limit estimation based on dissipated energy

1. The measured results are divided into two groups, and an approximate straight line is generated for each group. The residual sum of squares for each group is calculated about some grouping. The grouping which minimizes the residual sum of squares is decided as the suitable one.
2. The point where the two approximate straight lines cross is determined as the estimated fatigue limit.

23.3 Experimental Set Up

The single bead-on welding is applied to the JIS type SM490YA steel plate and the specimens were cut from the welded plate as shown in Fig. 23.3. The welded part is located on the center of specimen. Mechanical and thermal properties of SM490YA are shown in Table 23.1. Full reversed cyclic axis loading ($R = -1$) with a frequency of 5 Hz was applied to the specimen by the electrohydraulic servo fatigue testing machine.

Temperature change on the specimen surface was measured by infrared thermography with a MCT array detector. When the temperature of object which moves periodically was measured, the pretended temperature change was observed at the edge of the object, called edge effect. In this study, the position correction image processing was applied to the measurement of temperature for obtaining the appropriate image around the weld toe. Dissipated energy are obtained from measured temperature change by the lock-in algorism using Fourier analysis with phase information of dissipated energy (phase $2f$ lock-in infrared method) [8]. The double frequency component of temperature change on the specimen includes the influence

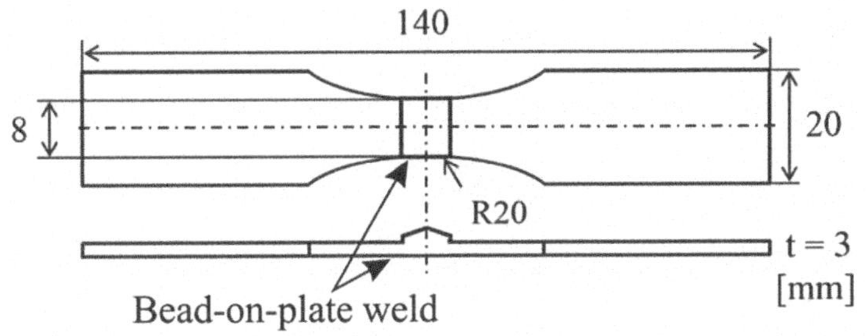

Fig. 23.3 Schematic illustration of specimen with bead-on-plate weld

Table 23.1 Mechanical and thermal properties of SM490YA

Yield stress, σ_Y [MPa]	Tensional stress, σ_B [MPa]	specific heat, c[J/(kgK)]	Density, ρ [kg/m^3]
394	535	477	7850

of the energy dissipation and harmonic vibration of the fatigue testing machine. The phase difference between the temperature change due to the energy dissipation and the wave with double frequency of thermoelastic temperature change Δθ is a specific value [8]. The phase $2f$ lock-in infrared method utilizes a specific phase of the dissipated energy, as following equation;

$$\Delta\widehat{T_D} = \frac{2}{N}\sum_{t=0}^{N} T(t)\cdot \sin\left\{2\cdot\left(2\pi\cdot t\cdot\frac{f_{\text{load}}}{f_{\text{meas}}} - \left(\theta_E + \Delta\widehat{\theta}\right)\right)\right\} \tag{23.2}$$

where f_{load} and f_{meas} are the load frequency and the measurement frequency of the thermal camera, respectively. A correlation with a negative value is set to zero. The phase $2f$ lock-in infrared method is effective for removing the noise component such as the thermoelastic temperature change due to the harmonic vibration of fatigue testing machine.

23.4 Result and Discussion

S-N curve for welded specimen is shown in Fig. 23.4. The fatigue limit are determined 160~220MP from Fig. 23.4. The distribution of dissipated energy at σa = 300 MPa during the stair-case like stress level test is shown in Fig. 23.5. The high dissipated energy appeared along the weld toe, and local concentration of dissipated energy can be observed at the upper and right side area of weld toe. After the stair-case like stress level test, the constant stress amplitude fatigue test was conducted for this specimen. The principal crack initiated from the upper and right weld toe and propagated along the weld toe. The two evaluation areas (10 pixels × 10 pixels) were selected in Fig. 23.5, Area 1 is the local high dissipated energy point, and Area 2 is located at the center of weld toe. The changes of dissipated energy in stair-case like stress level test are shown in Fig. 23.6. The change of dissipated energy in both evaluation area shows sharply increase from around 200 MPa. The solid line and dot line in Fig. 23.6 indicates the fitting line for Area 1 and Area 2, respectively. The value and increase rate of dissipated energy for Area 1 are larger than that for Area 2. The estimated fatigue limit is shown in Table 23.2. The estimated fatigue limit for Area 2 are determined from the boundary value of two dissipated energy groups in the estimation process described at previous section. The estimated fatigue limit for Area 1 coincides with fatigue limit obtained from S-N curve. High dissipated energy point in the early period of fatigue test before crack initiation means that the crack is more likely to initiate [9]. Therefore, it is considered that the fatigue limit can be estimated based on the change of dissipated energy at the point where high dissipated energy is observed at the end of stair case-like stress level test.

23.5 Conclusion

The fatigue limit estimation based on the dissipated energy measurement was applied to the fatigue strength evaluation of single bead-on-plate weld. High dissipated energy in the stair-case like stress level test was measured at the weld toe. The point where local concentration of high dissipated energy was measured along the weld toe coincided with the crack initiated point in the subsequent fatigue test. The fatigue limit was estimated from the change of dissipated energy at the local high dissipated energy point. The estimated fatigue limit shows the value near to the fatigue limit obtained from the *S-N* curves.

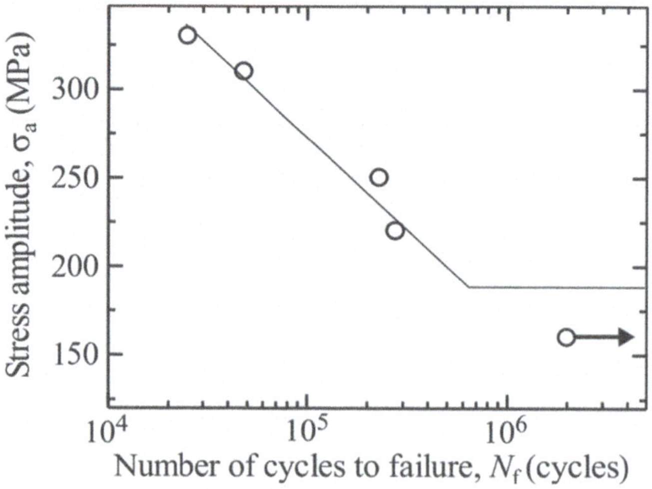

Fig. 23.4 *S-N* curves

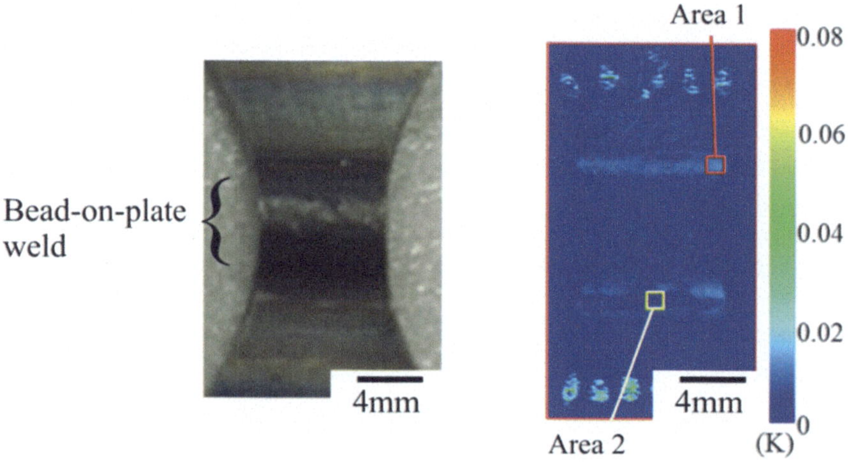

Fig. 23.5 Photograph of specimen and distribution of dissipated energy. (Stair-case like stress level test, $\sigma_a = 300$ MPa). (**a**) Photograph of welded specimen. (**b**) Dissipated energy distribution

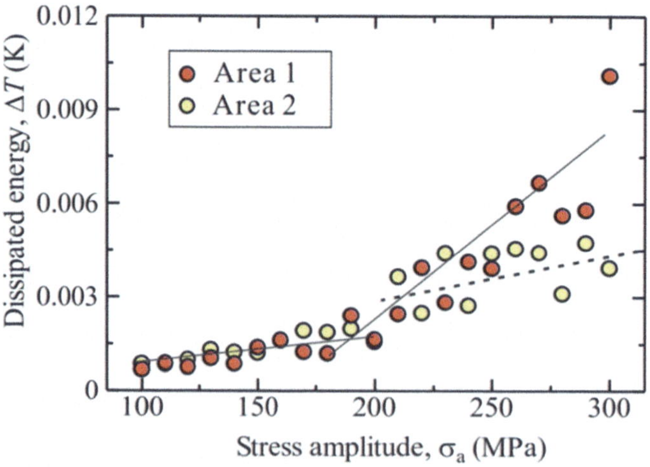

Fig. 23.6 Relationship between stress amplitude and dissipated energy

Table 23.2 Estimated fatigue limit based on the dissipated energy

Area 1	Area 2	*S-N* curves
194 MPa	205 MPa	160–220 MPa

References

1. La Rosa, G., Risitano, A.: Thermographic methodology for rapid determination of the fatigue limit of materials and mechanical components. Int. J. Fatigue. **22**, 65–73 (2000)
2. Krapez, J.-C., Pacou, D.: Thermography detection of damage initiation during fatigue tests. Proc. SPIE. **4710**, 435–449 (2002)
3. Luong, M.P.: Infrared thermography of fatigue in metals. Proc. SPIE. **1682**, 222–233 (1992)
4. Shiozawa, D., Inaba, K., Akai, A., Sakagami, T.: Experimental study of relationship between energy dissipation and fatigue damage from observation of slip band by atomic force microscope. Adv. Mater. Res. **891-892**, 606–611 (2014)
5. Akai, A., Shiozawa, D., Sakagami, T.: Fatigue limit evaluation for austenitic stainless steel. J. Soc. Mater. Sci. Japan. **61**(12), 953–959 (2012)
6. Palumbo, D., Galietti, U.: Characterisation of steel welded joints by infrared thermographic methods. Quant. Infrared Thermogr. J. **11**(1), 29–42 (2012)
7. Irie, Y., Inoue, H., Mori, T., Takao, M.: Evaluation of fatigue limit of notched specimen by measurement of dissipated energy. Trans. Japan. Soc. Mech. Eng. Ser. A. **76**(764), 410–412 (2010)
8. Shiozawa, D., Inagawa, T., Washio, T., Sakagami, T.: Accuracy improvement in dissipated energy measurement by using phase information. Meas. Sci. Technol. **28**, 044004 (2017)
9. Akai, A., Inaba, K., Shiozawa, D., Sakagami, T.: Estimation of fatigue crack initiation location based on dissipated energy measurement. J. Soc. Mater. Sci. Japan. **64**(8), 668–674 (2015)

Chapter 24
Thermoelastic Measurement Techniques Enabled by Self-reference

Bradley R. Boyce and Jon R. Lesniak

Abstract Until recently thermoelastic stress analysis (TSA) imaging required correlation between an electronic signal related to the stress in a target structure, and the thermal signal acquired in successive IR images. With the implementation of a new method dubbed "self-reference" the requirement for the electronic signal has been eliminated. In the self-reference modality, correlation is made between the temporal variation in the signal from a region-of-interest in the image, to each of the pixels in the image. This simple change of correlation parameters has had a dramatic effect on the applicability of TSA and the ease of use of TSA instruments. A review of a range of applications in the light of this change indicates a new level of applicability of the TSA technique for stress measurement.

Keywords Thermoelasticity · Self reference · Stress · Constant amplitude · Variable amplitude

24.1 Overview

A change of reference signal source for the processing of infrared images brings a long list of new and extended capabilities, that enhance and applicability of TSA. Among the new applications and features enabled or enhanced by self-reference are: impulse analysis, repeated processing of loading histories, greater applicability of TSA to higher and more damaging loads, facilitation of crack detection techniques, and easier implementation of TSA for random and vibration loadings. Additionally, the self-reference processing makes the TSA technique easier to grasp, and the interpretation of results simpler. The change of reference signal source also impacts hardware issues: self-reference enables the use of less expensive cameras for TSA, improves portability, reduces the cost of the frame-grabber and camera. The ability to use non-deterministic frame grabbers is no longer an issue, opening the door to video over ethernet and even networked cameras. Off the shelf wireless connections to cameras and video recording equipment further extends the options for system configuration.

24.2 TSA Signal Processing

Traditionally the formation of a TSA image is calculated by correlating infrared pixel values to a reference signal through a least squares method [1, 2]. The formula that describes the TSA image, I_{tsa} is shown in Eq. 24.1.

$$I_{tsa} = \frac{N\Sigma\,(S * F) - \Sigma S * \Sigma F}{N\Sigma F^2 - (\Sigma F)^2} \tag{24.1}$$

Where: N is the number of captured images.
S is the sampled image
F is the reference value

This formula is at the heart of most modern TSA instruments. It is worth noticing how the formula might be implemented. Say that a 2-second-long series of images is collected. A typical camera framerate would be 100 frames/second, so 200

B. R. Boyce (✉) · J. R. Lesniak
Stress Photonics Inc., Madison, WI, USA
e-mail: bboyce@stressphotonics.com

A. Baldi et al. (eds.), *Residual Stress, Thermomechanics & Infrared Imaging, Hybrid Techniques and Inverse Problems, Volume 7*, Conference Proceedings of the Society for Experimental Mechanics Series, https://doi.org/10.1007/978-3-319-95074-7_24

images. The images are processed, and a resulting image is produced. This "accumulated" image is a view of the relative stresses in the structure. Averaging this image with others, collected subsequently, will improve the quality of the result.

24.3 Variable Amplitude Loading

For measurements made during variable amplitude loading there are several special conditions that must be met. First, the reference signal must be directly or inversely proportional to the loading of the structure. There must be no phase shift between the stressing of the structure and the reference signal (in-sync). Finding a reference signal with the necessary proportionality and without phase shift may be difficult. For instance; if the loading of a structure is caused by multiple uncorrelated loads, or if there is a dynamic response in the structure or in the loading fixture. One solution to this problem is using the signal from a strain gage installed near the area of interest.

The strain gage solution has shortcomings in that there might be more than one area of interest or more than one reference signal of interest. Since most TSA systems allow for only a single channel of reference single the loading sequence might need to be run several times. This requirement can eliminate the TSA as a measurement technique for applications where only one highly damaging load application is the target of the investigation; for instance, an automotive "pot hole" test. An alternative to the strain gage approach is to use a virtual strain gage to generate the reference signal (also known as Self Reference). By virtual strain gage it is meant, the signal derived from the average of the pixel values from a small area of each infrared image. Then if a recording of the infrared images is stored, the virtual strain gage can be applied in different locations on a series of playbacks and the results for a set of potentially uncorrelated references can be calculated.

There is another advantage to the self-reference method that is not obvious from this explanation. With an electronic reference signal is important to sample signal simultaneous with the camera exposure. Depending on the frequency content of the signal relative to the exposure period it might be appropriate to average the electronic signal over the period of the exposure. In any case, a coordination between the camera framing and signal sampling must be maintained. This is inherent in the self-reference method but requires a frame sync signal from the camera in the case of an electronic reference signal.

24.4 Constant Amplitude Loading

In fact, there is a common instance of multichannel reference in TSA. It is common to use a sine and cosine wave pair as the references for a structure that is loaded at a constant amplitude. The resulting sine and cosine correlated TSA images can be combined to represent the magnitude and phase of the stresses in the structure. By phase it is meant the relative timing between the reference wave and the actual loading. The phase information can be used in two ways: as a tool to make sure that maximums are not missed, and as a tool to infer information about thermal conduction, plasticity, and other phenomena.

For constant amplitude loading a virtual strain gage approach can also be used. In this instance the average of values in a small subset of each incoming image are processed and a history of the values stored. A sinewave is fit to these values and the measured frequency used to synthesize sine and cosine references. Using the "accumulation" period described above is convenient. During each accumulation period the synthesized sinewave can be used from the previous accumulation, and then, before the accumulated image is averaged in with the previous images, the synthesized sine wave can be verified by again fitting the measured virtual strain gage frequency and phase for the current accumulation period.

When using constant amplitude loading the reference source is required to be synchronized but not proportional to the stresses in the structure. Synchronization can come from selecting a virtual signal that is not strain related, but is, for instance, displacement related. In TSA small displacements of the target structure can cause edges of the target to stand out in the TSA image, because the edge may be chopping the signal from a dissimilar temperature located behind the structure, thereby causing a large signal known as an edge-effect. Edge-effects are usually large signals often found somewhere in the camera view. Using the edge-effect signal to generate the synchronized sine-cosine wave pair is a useful technique.

24.5 Triggered TSA

The use of edge-effect signaling can be extended to enable additional capabilities. The signal from an ROI can be used for triggering the start and/or end of processing or recording. Given a processing framework that includes a circular buffer of recorded images a trigger can occur before or after the start of a recording. The ROI used for triggering need not be the same as the ROI used as a virtual strain gage for a reference. Furthermore, an all-inclusive recording can be made to assure the capture of a critical event. Then in a post-processing session there is flexibility in selecting an appropriate ROI for triggering and an appropriate ROI a reference signal.

24.6 Enabled Technology

The implementation of the self-reference modality is simple but powerful. We have already discussed some important implications for the applicability of TSA for impulse loading. But another important point is that, a loading that can only be applied once because of its destructive nature is now within the comfort range of TSA.

Using a virtual strain gage reduces the constraint of finding a synchronized and proportional reference signal. Additionally, the quality of the reference signal in terms of timing becomes a non-issue. This broadens the use of TSA in the areas of dynamic loading, but also has impact on the selection measurement hardware.

Since an electronic reference signal is not needed TSA systems can move from desktop computers with traditional data acquisition hardware to a laptop computer. Portability is enhanced by this and by implementation of GigE Vision in IR cameras. A GigE Vision capable IR camera uses ethernet for the transport of video. Consequently, the ethernet port of a computer can be used as the frame grabber. The convenience and reduced cost of the GigE Vision video interface that is enabled by the self-reference since it is inherently in-sync with the video. Therefore, the non-deterministic nature of GigE Vision frame grabbers is not a problem. To take this a step further, GigE Vision cameras can also make use of inexpensive wireless links and network architectures.

Without self-reference is it important to have a known timing between the sampling of the electronic reference signal, and the framing of the camera. The timing requirement dictates the camera must have a frame sync output or input. This frame sync feature is less likely to be available on lower cost cameras. Another way that self-reference lowers the cost of the camera is by allowing for the use of microbolometer cameras that are substantially lower cost. The trick of using an edge-effect ROI for generating the timing needed for a constant amplitude reference signal assures that a strong signal is available even when the observed area might have a low signal-to-noise for the TSA imaging.

The idea of using multiple IR cameras to monitor a structure undergoing a long-term test has been thwarted by the cost of cameras. Microbolometer cameras are much lower cost, but to implement them without self-reference requires a frame sync signal and a deterministic frame grabber. With these constraints eliminated by self-reference, multi-camera application can be considered again.

24.7 Conclusion

Students of experimental mechanics and potential users of TSA often have a hard time with the concept of using a reference signal for image processing. Then without a solid understanding of the problems associated with getting a correlation between a reference signal transducer and the pixel values, TSA users can easily misunderstand results. The use of the self-reference modality eliminates the necessity of considering the quality and character of the reference signal. We find that correlating the pixels of an image to a small group of pixels from within the same image is a simplifying and dramatically enabling step forward.

References

1. Lesniak, J.R., Boyce, B.R.: A High-Speed Differential Thermographic Camera, SEM Spring Conference, June 1995
2. Lesniak, J.R., Boyce, B.R., Howenwarter, G.: Thermoelastic Measurement Under Random Loading, SEM Spring Conference, June 1998

The manufacturer's authorised representative in the EU is Springer
Nature Customer Service Centre GmbH, Europaplatz 3, 69115 Heidelberg,
Germany. If you have any concerns regarding our products, please
contact ProductSafety@springernature.com

Printed and bound by CPI Group (UK) Ltd, Croydon, CR0 4YY
06/10/2026
02228820-0003